AF696104

PUBLIC POLICY
IN INDUSTRIAL GROWTH

This is a volume in the Arno Press collection

DISSERTATIONS
IN
EUROPEAN ECONOMIC HISTORY

Advisory Editor
Stuart Bruchey

Research Associate
Eleanor Bruchey

Editorial Board
Rondo Cameron
Carlo Cipolla
François Crouzet
Wolfram Fischer
Ira A. Glazier
Peter Mathias
William N. Parker

See last pages of this volume for a complete list of titles.

PUBLIC POLICY IN INDUSTRIAL GROWTH

The Case of Ruhr Mining Region 1776-1865

Manfred Dieter Jankowski

With a New Preface and Appendix

ARNO PRESS
A New York Times Company
New York / 1977

Editorial Supervision: ANDREA HICKS

First publication 1977 by Arno Press, Inc.

Copyright © 1977 by Manfred Jankowski

DISSERTATIONS IN EUROPEAN ECONOMIC HISTORY
ISBN for complete set: 0-405-10773-0
See last pages of this volume for titles.

Manufactured in the United States of America

Library of Congress Cataloging in Publication Data

Jankowski, Manfred Dieter.
Public policy in industrial growth.

(Dissertations in European economic history)
A revision of the author's thesis, University of Wisconsin, 1969.
Bibliography: p.
1. Coal mines and mining--Germany--Ruhr Valley--History. 2. Industry and state--Germany--Ruhr Valley--History. I. Title. II. Series.
HD9553.7.R8J36 1977 338.2'7'2094355 77-77175
ISBN 0-405-10788-9

PUBLIC POLICY IN INDUSTRIAL GROWTH:

The Case of Ruhr Mining Region, 1766-1865

BY

MANFRED DIETER JANKOWSKI, PH.D.

UNIVERSITY OF WISONSIN, 1969

PREFACE

During the hundred years from 1766 to 1865 the Ruhr mining region, the modern heart of the Rhenish-Westphalian industrial region, expanded with development of the coal field. In the same period, public policy in the form of Prussian and French state intervention had a significant impact on the direction and speed of the region's economic growth. Specifically, changes in mining and other laws and their administration affected the Ruhr region and its mining industry. In turn, changes in the regional economy, transportation, technology, business organization, and entrepreneurial activities affected law and state policy. During the century under discussion state policy moved from one of strong control over mining and related industries with severely restricted private enterprise to one of free enterprise with a minimum of state regulations.

This work consists of two parts, a dissertation presented in 1969 as part of the Ph.D. requirements of the Economic History Program of the University of Wisconsin, and a "Statistical Appendix" based on archival research since then. Quite naturally I wish to acknowledge the important stimulus from the excellent teachers of the program and particular its then director and my dissertation advisor Rondo E. Cameron. I also wish to thank the University of Wisconsin for timely granting of a Dissertation Completion Fellowship during the summer of 1968 while already teaching at Northern Michigan University.

The "Statistical Appendix" reflects only part of the data from extensive archival research in Germany. I appreciate the financial assistance of Loyola University of Chicago, allowing research in Germany in 1970. I also am most grateful for the generous Grant-in-Aid (D.A.A.D. 1971) from the Federal Republic of Germany permitting the completion of my work in archives in Münster, Düsseldorf, Dortmund and the Mining Library in Essen. Some of the results of this research were included in an article, "Law, Economic Policy, and Private Enterprise: The Case of the Early Ruhr Mining Region, 1766-1865," *The Journal of European Economic History*, II (Winter, 1973).

Finally, I owe many acknowledgements for help I received from many individuals in the research, microfilming, data processing, computing, compiling and typing of voluminous data reflected in the appendix. Of course, my special thanks goes to my wife for her great patience and diligence.

Manfred D.Jankowski

TABLE OF CONTENTS

Page

INTRODUCTION 1

CHAPTER I THE *DIREKTIONSPRINZIP* AS A PRUSSIAN POLICY BEFORE 1851: HISTORY AND ADMINISTRATION 8

A. Prussian Economic Policy: Silesia and the Ruhr 8

B. History of the *Direktionsprinzip* to 1766 16

C. The Application of the *Direktionsprinzip* under Freiherr vom Stein . 22

D. The Administration of the *Direktionsprinzip*, 1766-1851 . . . 26

CHAPTER II OBJECTIVES AND REGULATIONS UNDER THE *DIREKTIONSPRINZIP*, 1766-1850 43

A. Production Controls 44

B. Managerial and Administrative Controls 52

C. Market Controls 57

CHAPTER III THE STATE AND TRANSPORTATION IN THE RUHR 78

A. Introduction 78

B. The State and River and Road Improvement 81

1. The Ruhr River 81

2. The Rhine River 90

3. The Lippe and Emsche Rivers . . 94

4. Roads 97

C. The State and Railway Development . 101

1. General Prussian Policy 102

2. The *Direktionsprinzip*: Coal Railways 106

3. Railroad Development 112

CHAPTER IV STATE ADMINISTRATION AND PRIVATE ENTERPRISE: CONFLICT AND GROWTH IN THE FIRST HALF OF THE NINETEENTH CENTURY 127

Introduction 127

A. Production Controls and the Schaft Mine 130
B. Marketing and Market Controls 139
1. Marketing 139
2. Market Controls: Price System . . 153
C. Business Organization and Finance . . 168
1. The Mining Company and Joint-stock Company 168
2. The Tax System Under the <u>Direktionsprinzip</u> 175
D. Mining Law and Its Administration . . 179

CHAPTER V LEGAL REFORMS OF THE <u>DIREKTIONSPRINZIP</u> . . 192

A. General Prussian Reforms 192
B. Unsuccessful Attempts at Mining Law Reform, 1826-1851 203
C. Reform of the <u>Direktionsprinzip</u>: Reform Laws, 1851-1865 207
D. Minister Von der Heydt and the Opposition to the Reform Laws 218

CHAPTER VI THE REFORMED <u>DIREKTIONSPRINZIP</u> AND ECONOMIC CHANGE IN THE MINING INDUSTRY: 1851-1865 232

A. The Growth of the Ruhr Coal Industry and Large Operations, 1851-1865 . . . 232
B. The Effect of the Reform and the Growth of the Joint-stock Company . . 236
1. The Tax Reform 236
2. The Economic Effects of the Shareholder Law 252
3. The Joint-stock Company 256
C. The Marketing and Consumption of Coal 261

CHAPTER VII THE <u>DIREKTIONSPRINZIP</u> IN HISTORICAL PERSPECTIVE 275

A. The Codification of 1865: The End of the <u>Direktionsprinzip</u> 275
B. The <u>Direktionsprinzip</u> in Perspective . 281

BIBLIOGRAPHY 286

LIST OF TABLES

Table		Page
2.0	Production of Coal: 1764-1787	68
2.1	Production, Mines, Employment, Value of Sales: 1788-1850	69
2.2	1788-1850 Averages: Production Per Mine, Employment Per Mine, Production Per Worker, and Price	72
4.0	Early Steam Engines and Shaft Mines in the Essen Region	140
4.1	Steam Engines in the Essen Region: 1821-27 .	141
4.2	Share of Largest Mines of 1850 Output	142
4.3	Production of the Ten Largest Shaft Mines in 1850	142
4.4	First Shafts Penetrating the Water-Bearing Rock (<u>Mergeldecke</u>)	143
4.5	Mining Activity in the Iron Ore (Blackband) Industry: 1847-1852	154
4.6	Coal Production: 1830-1840	155
4.7	The First Joint-stock Companies in the Ruhr Region: 1840-1853	176
4.8	Tax Table, Based on Five Year Average Value of Production 1843-47	180
4.9	Tax Table, Percentages of Table I	181
6.0	Production, Employment, Price, Mines: 1850-1865	237
6.1	Average Production and Employment Per Mine and Productivity per Work	238

Table		Page
6.2	Index Numbers of Production, Mines, Employment and Price (1850 = 100)	239
6.3	Yearly Percentages Changes, 1850-1865	240
6.4	Three Year Average Rates of Change: 1851-1865	241
6.5	Number of Mines and Size Classes	242
6.6	Production of Mines over 50,000 Tons	243
6.7	Mines Producing Over 50,000 Tons	244
6.8	Output Per Worker, 1850-1865	245
6.9	Steam Engines and Their Horse Power in the Essen, Bochum Regions	246
6.10	Production and Price: 1788-1803, 1835-1850, 1850-1865	247
6.11	Mines and Employment: 1788-1803, 1835-1850, 1850-1865	249
6.12	Output and Employment Per Mine, Output Per Worker: 1788-1803, 1835-1850, 1850-1865	250
6.13	Taxes: 1850-1866	253
6.14	35 Newly Established Joint-stock Companies: 1854-1857	262
6.15	Activity in Mining 1852-1860	263
6.16	Coke Production and Use in Blast Furnaces 1850-65	267
6.17	Coke and Pig Iron Production: 1850-1865	268
6.18	Coal Consumed by Iron and Steel Works, 1851-1856 (Per Cent of Total Coal Production)	269
6.19	Sale of Coal Using the Ruhr, Railroad and Other Means: 1851-1861 (Per Cent)	269
6.20	Coal Transport on the Cologne-Minden and Berg-Mark Railways: 1854-1861	270

Table Page

6.21 Transport of Coal on the Cologne-Minden and Berg-Mark Railways and Their End Points: 1855-57 270

6.22 Steamships on the Lower Rhine: 1850's 271

6.23 Export of Coal from the Ports of Duisburg and Ruhrort to the Rhine, 1850-65 272

6.24 Water Transport of Coal from Ruhrort/Duisburg Ports on the Rhine: 1851-60 272

INTRODUCTION

The subject of this dissertation is the relation of Prussian mining law to the economic development of the Ruhr mining region during the century from 1766 to 1865. Specifically, it is concerned with the reciprocal relationship between changes in the mining law and its administration as well as the changes in economic activity, business organization, entrepreneurial motivation, transportation and technology.

Definition of the Ruhr region itself presents an initial difficulty. From the point of view of economic history the Ruhr mining region constitutes a unified area for analysis. Yet the factors considered in this analysis require us to see this region from various points of view. The region, therefore, must be described, though it cannot be absolutely defined from a geological, a geographical or a political perspective.

Physically, the entire Ruhr coal field extends from the Rhine River about sixty-five miles from west to east. The coal measures rise to the surface in the southern part of the field, but toward the north dips under cover of secondary water-bearing rock. The southern limit of the coal measure lies in some places north, in others south of the river Ruhr. For much of its lower course the river flows across an exposed field and the coal measures outcrop on the valley side as well as in the valleys of the numerous small tributaries. Consequently this was the area of early coal mining activity. Gradually mining became more difficult because the coal lay beneath the cover of later strata, and as these were to a high degree water-bearing, its extraction was difficult and costly, requiring a more advanced technology. The coking coal, which lay beneath the water level, was reached in the late 1830's. Furthermore, in the 1840's a local supply of iron ore similar to the English and Scottish Blackband ores was found.

Geographically, the Ruhr mining region expanded with the historical development of the coal field, developing from the south to the north. The oldest and southernmost area is the area around the Ruhr Valley. Most of this area is south of the Ruhr River. It begins below Mühlheim and includes the relatively small river cities. This was the area in which coal mining of sorts had taken place for hundreds of years, at least since the thirteenth century. It was easy to mine and often required no more than a shovel and a strong back. Until the 1820's nearly all of the Ruhr coal was mined here. Immediately north of the Ruhr Valley itself is the Hellweg ridge. This is the area where shaft mining began in the 1830's and where the iron industry eventually settled.

In an economic sense, the Ruhr industrial region, which derived its importance from its coal resources, may be thought of as bounded on the north and west by the Lippe and Rhine Rivers, both important transportation routes; in the east by the city of Dortmund and on the south by the Wupper Valley, where the cities of Barmen and Elberfeld provided an early and important market.

Politically, the major portion of the Ruhr region fell to Prussia by the Peace of Westphalia in 1648, but the entire Rhenish-Westphalian region, of which the Ruhr was the heart, remained geographically separated from the rest of Prussia until 1866. Until 1803, when Prussia acquired the ecclesiastical territories of Essen and Werden, these two territories separated Prussia's two largest and most valuable possessions in western Germany: The Duchy of Cleve and the County of Mark. Except for the period of French occupation from 1807 to 1815, the Ruhr region was permanently unified under Prussia after 1803.

For the purposes of this dissertation, the Ruhr region can be adequately defined only as an "economic development region," where the state attempted to encourage the economic growth of the region. The intervention of the Prussian state in the mining industry was expressed in mining and related laws. Between the 1760's and the 1860's state policy moved from one of strong state control over the mining industry with severely restricted private enterprise to one of free enterprise with a minimum of state controls. During most of this period Prussian policy toward the Ruhr region and its mining industry was characterized by what is called the _Direktionsprinzip_. The _Direktionsprinzip_ (principle of direction) was both a legal system and an economic policy. For the Prussian part of the Ruhr region, the _Direktionsprinzip_, in its narrow sense, referred to and was defined in the 1766 Cleve-Mark Mining Codification. But in a wider sense the term designates the government's overall economic policy toward the Ruhr region and the mining industry.

The period from 1766 to 1815 saw the full legal and economic implementation of the _Direktionsprinzip_. During this period the state assumed a leading role in the economic development of the Ruhr mining industry. The _Direktionsprinzip_ became the basis for a coordinated economic policy for the entire Ruhr region. During this period the state emphasized the development of policies, programs and actions to move the region from where it was toward predetermined economic objectives.

The period between 1815 and 1851 was characterized by the tension created between the state policy and actual economic and social conditions in the Ruhr mining region. While the economic conditions changed, the _Direktionsprinzip_ remained unchanged. The following period from 1851 to 1865 introduced a series of reforms which changed the established policy of the _Direktionsprinzip_, thereby freeing private enterprise from strict state control. This period of reform ended with the mining codification of 1865 which embodied a more comprehensive state policy toward the mining industry.

This study is concerned with the reasons for changes in both economic conditions and policy. Chapter I offers an exposition of the *Direktionsprinzip* as the official policy of the eighteenth century Prussian State and describes the conditions which called it into being, culminating in the Cleve-Mark Mining Codification of 1766. Chapter II describes the economic of objectives of the *Direktionsprinzip* and the methods of implementing them. Chapter III analyzes the state policy toward transportation and its effect on the Ruhr mining region. Chapter IV evaluates the overall effectiveness of the original conception of the *Direktionsprinzip*, especially in the light of changing conditions during the first half of the nineteenth century. Chapter V describes the reform of the *Direktionsprinzip* under the pressure of changing economic conditions, which lead to the reforms of the 1850's. Chapter VI considers the economic growth of the Ruhr region under the reformed *Direktionsprinzip* after 1851. The concluding chapter analyzes the General Prussian Mining Codification of 1865, which ended the *Direktionsprinzip*, and evaluates the effectiveness of the established state policy in the perspective of the hundred years of its operation.

THE RUHR IN THE MIDDLE OF THE NINETEENTH CENTURY

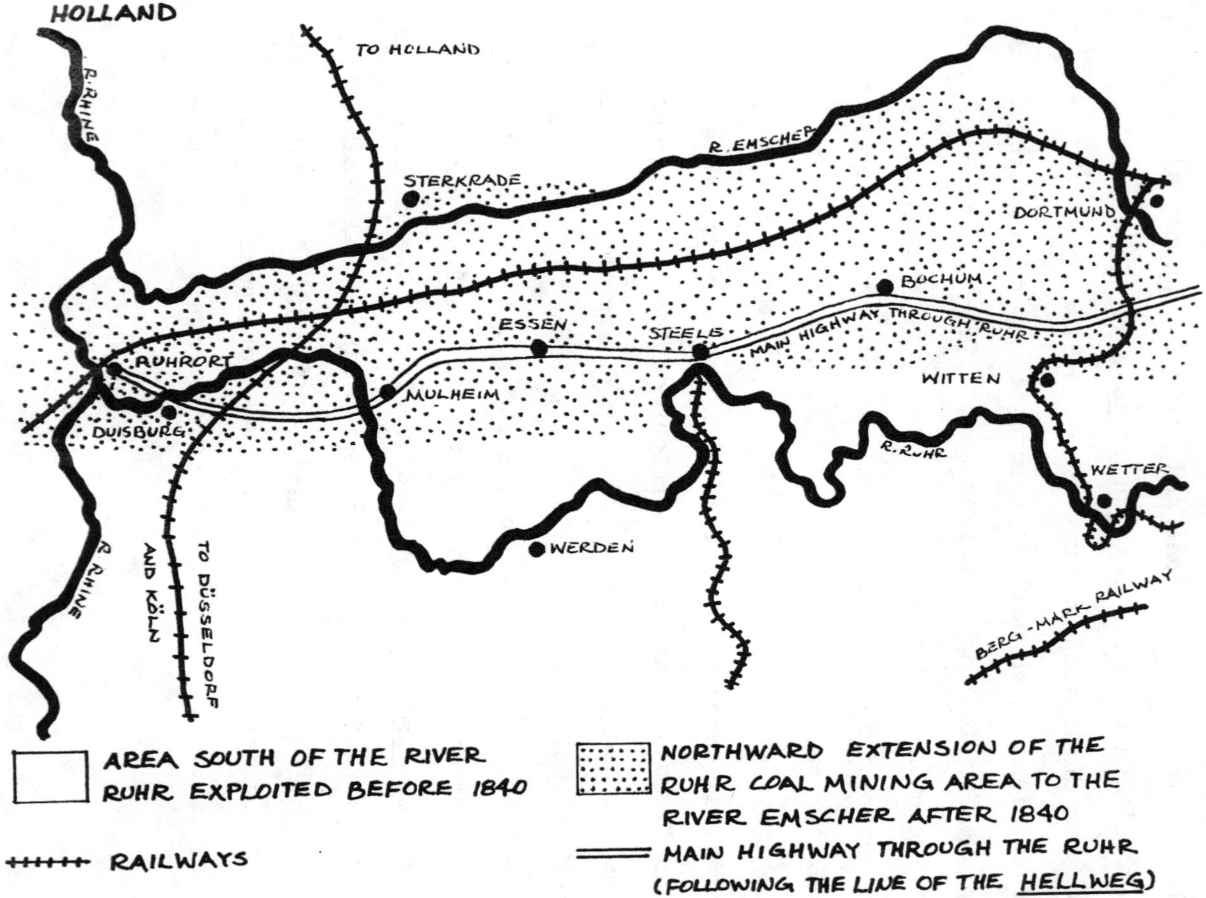

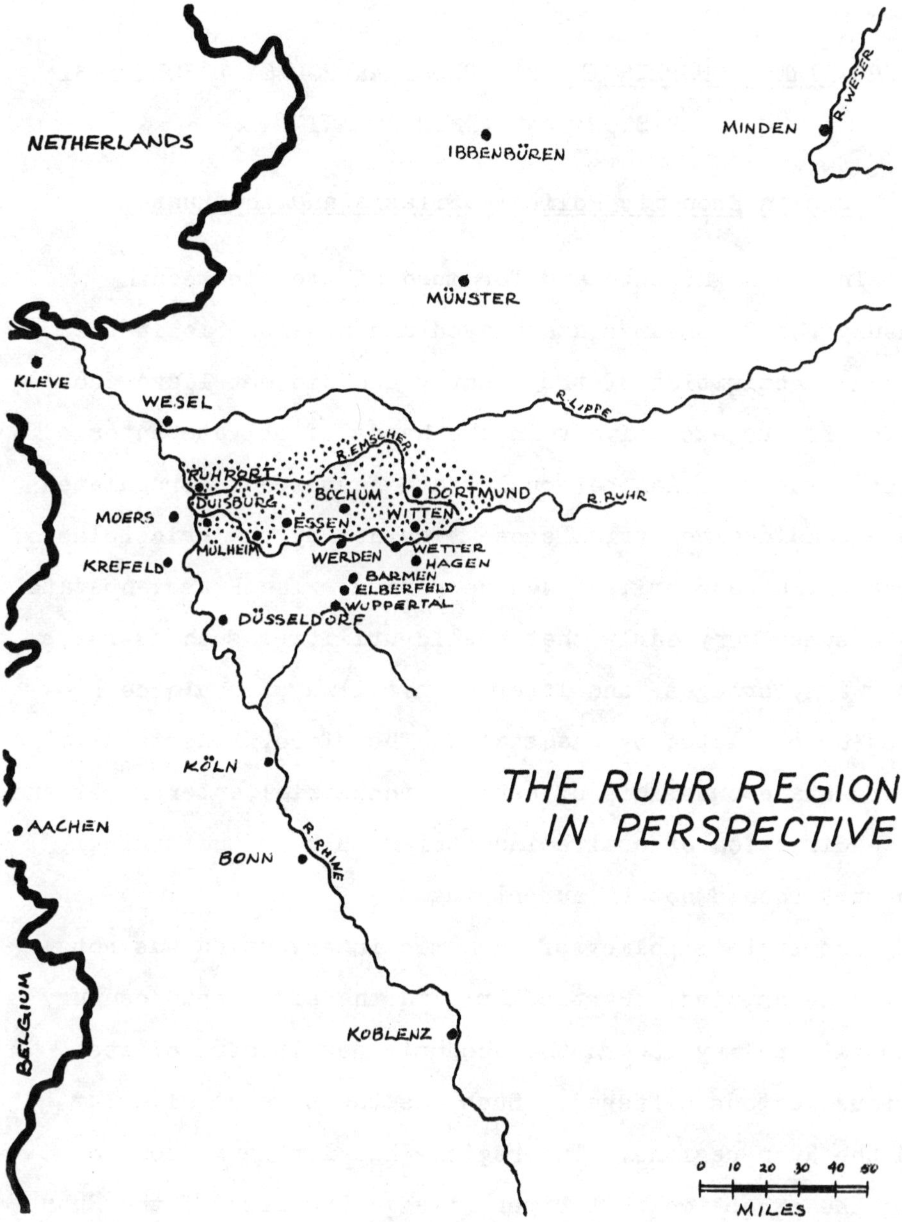

THE RUHR REGION IN PERSPECTIVE

CHAPTER I

THE *DIREKTIONSPRINZIP* AS A PRUSSIAN POLICY BEFORE 1851: HISTORY AND ADMINISTRATION

A. Prussian Economic Policy: Silesia and the Ruhr

In the eighteenth and for much of the nineteenth century the Prussian state played a dominant role in the economic activities of the country and did not leave economic affairs exclusively in the hands of private enterprise. The Prussian paternal state fostered and regulated the expanding industrial economy guided by the principles of mercantilism and enlightened despotism. The Prussian state had assumed very early that public utilities such as roads, waterways, bridges, and later even railways should be provided or regulated by the state. Therefore, Prussia's policy toward state ownership of certain industrial enterprises and state direction of entire industries such as the mining industry should not be surprising.

The state's policy of economic intervention was not uniformly applied. Particularly in the eighteenth century Prussia's policy toward the economic development of its various regions differed. Such was the case of Silesia[1] and the Ruhr regions. The regions had little in common with the exception that Upper Silesia and most of the Ruhr region were situated on coal fields. State policy was adjusted to the different geographical, political, and economic conditions of these regions.

In the eighteenth century, in both cases, the _Direktionsprinzip_ was the guiding principle for a specific regional economic policy. In Silesia, but particularly in Upper Silesia, it became the basis for an active nationalization policy in the mining industry. In the Prussian part of the Ruhr region, the _Direktionsprinzip_, although specifically a system of regulation and control over the coal mining industry in the County of Mark, became after 1766 the basis of a general economic policy by the Prussian state toward the western territories. Particularly in the last quarter of the eighteenth century, it provided the basis for a co-ordinated policy of the state to stimulate the economic growth of these territories.

Before the Prussian acquisition of Silesia mining rights, under the Emperor Rudolph's mining law of 1577, belonged to the property owners. These large estate holders were deeply in debt and had shown little interest in the efficient exploitation of the iron and coal in the region. There existed a small iron industry and a coal industry which was extremely backward due to the existence of an easily accessible local wood supply. These backward conditions remained until Frederick the Great enforced the _Direktionsprinzip_ in iron and coal mining and allowed state ownership, especially of ironworks.[2]

In the second half of the eighteenth century, the state followed in Silesia a policy of both encouraging existing private enterprise as well as establishing state-owned enterprises.[3] It did not intentionally suppress private initiative and ownership, but, in the absence of these, created a viable industry which would guarantee the state its military supplies. In Silesia, therefore, both state and private ownership existed. The state owned ironworks and some ore and coal mines. While in Lower Silesia nearly all the ore and coal mines were privately owned but under direct state control, in Upper Silesia state ownership was predominant.[4]

In 1780, when Von der Reden became the leading mining official for Silesia, he found extremely backward conditions. In all of Silesia there were only 27 coal mines with a total employment of 265 men and a total production of 126,000 tons, with the three largest mines producing nearly one third of the total. Only eight mines made a profit. In addition, there were only three iron ore mines and five other ore mines with a combined employment of 110 men. None made a profit. Finally, there were also two state owned ironworks, one at Malepane and the other at Kreuzburg.[5]

By 1791 under the capable leadership and direct supervision of the mining operations by Von der Reden, the total coal production from the privately owned mines in Lower

Silesia had reached 386,000 tons, or three times that of 1780. In Upper Silesia coal production from the mostly state-owned mines was 100,000 tons. Further, the state had increased its ownership of iron and other ore mines as well as iron works.[6] In general the state had assumed economic leadership in Silesia and ownership especially in Upper Silesia of the iron and coal industries.

Under this policy the eighteenth century Prussian state became the initiator of concentration and combination of the individually state-owned enterprises, as well as unifying the various production stages. This came about despite the fact that all the kings were in principle against state monopolization. Frederick the Great was against increasing state ownership, as were Frederick Wilhelm II and Frederick Wilhelm III.[7] Nevertheless, under this state leadership Silesia emerged by the end of the eighteenth century as a relatively advanced industrial region. The Ruhr coal industry was not able to catch up with Silesian coal production until the 1850's.

The Ruhr region presented a different case. While the mineral resources in Silesia were located far from large centers of population and the inadequacy of communication hampered industrial progress,[8] the Ruhr was more favorably situated in this respect. Its coal and iron products could be sent to the cities in the Rhine valley and even the Low Countries. However, even though transportation facilities had been improved during the eighteenth century, they left much to be desired. Furthermore, the Ruhr region in itself was splintered, and lay in a detached position from the central part of Prussia with which there was only limited communication.[9]

There was moreover a difference in the political and social structure of the two regions. Historically Silesia was seized in a single campaign and was successfully defended against Austrian attempts to recover the province. The scattered Prussian possessions in the western part of Germany were acquired piecemeal over a period of two hundred years. Politically, Silesia, like Prussia's territories east of the Elbe, had no organs of self-government. Consequently, state officials had little concern for local public opinion regarding most measures they introduced.[10] In the western territories the position was quite different. Provincial organs of local self-government had survived. Government officials had to justify their policy before the local estates (<u>Stände</u>). As Henderson points out, these provincial assemblies had preserved three important rights. They were called regularly, they approved the local budgets, and they discussed proposed laws. In addition, at a lower level there were in most county districts regular meetings of the nobles and the peasants to discuss local affairs and to appoint minor officials.[11]

"To secure smooth government of the western provinces,

even Frederick the Great had to adjust his policy to meet the needs of a system of administration which differed from that of the main part of the kingdom."[12] In the west, efficient government could not be secured simply by orders from Berlin. The effective administration involved not only the ministers and officials in Berlin but the officials on the spot and the local estates.[13] For these reasons the highest political representatives of the state in the western territories possessed a great deal more independent policy making power than normally would be the case. Even in the first half of the nineteenth century the Prussian government in Berlin continued to look with greater favor on the industrial center that was, if not appreciably nearer, at least more amenable to its supervision: Silesia.

During the eighteenth century the general Prussian policy was to make the western territories self-sufficient and an asset to the rest of Prussia. Especially under Frederick the Great the state during the second half of the eighteenth century implemented this policy goal by encouraging population growth, introducing agricultural reforms, stimulating regional industries, effectively introducing and enforcing the <u>Direktionsprinzip</u> in the Mark coal industry, providing capable state leadership, overcoming political obstacles, and initiating improvements in transportation.

In the Ruhr region the population after the Thirty Years War had actually decreased. Even by the second half of the eighteenth century, it was only slowly increasing. In the early eighteenth century under Frederick Wilhelm I the Prussian government had encouraged the young to settle in the newly acquired eastern territories. The drain of population was not offset by the immigration of French refugees, who went mostly to the neighboring Duchy of Berg. Only under Frederick the Great was the emigration stopped.[14] He thought it wiser that the western provinces should be built up so that they could prosper and supply the agricultural produce demanded within these territories. In the agriculturally deficient areas, such as the Duchy of Cleve, he wanted to encourage industrial undertakings.

Consequently a prohibition of emigration was gradually enforced. Discharged military personnel were to stay in the area. Further, the sons of farmers should be given a plot of land so that the farm communities, especially in the agriculturally rich Mark, would grow. By tradition, in the County of Mark, plots of land and individual households were not divisible among several children. Frederick the Great discouraged the practice of primogeniture. Further edicts followed even during the Seven Years' War. In 1764, colonists were ordered to make the wasteland agriculturally productive. In the same year the cities of Bochum, Unna and Soest, which were located in wheat rich areas, were ordered to establish new villages in their districts.[15]

In 1769, a "Publicandum for Immigrants" gave newcomers

who would settle in previously non-agriculturally productive land exemption from military service. They came under the special protection of the king. Not even the local government officials were to interfere in their activities.[16] Owing in part to these reforms and the gradual increase in population during the second half of the eighteenth century, the County of Mark began to export agricultural produce, especially to the nearby but politically separated, more industrial Prussian Duchy of Berg.

Although during the second half of the eighteenth century the major industry in the County of Mark continued to be agriculture, manufacturing activities also increased. The textile and woolen industries began to blossom. A glass industry started. The tobacco processing industry developed with the supply of American tobacco. By 1784, there were seven tobacco factories, all of them established since 1750.[17] In the eighteenth century Rasenstein became the basis for an infant iron industry in the Ruhr region. The location of this low grade iron ore and not the coal became the determining economic factor in the location of the first ironworks. This was the case of the Antonihütte near Starkrade in 1753 as well as the St. Michael Hütte established in 1730.[18] The development of this early iron industry in the Ruhr region was quite independent of the older iron industry in the southern part of the Mark, the Sauerland.

Under Frederick the Great the state urged the development

of the coal industry in the County of Mark. The 1766 Revised Cleve-Mark Mining Codification was to provide the basis for the growth of this industry. Furthermore, the Direktionsprinzip, although specifically established for the mining industry, was to provide the basis at least for the second half of the eighteenth century for a co-ordinated economic policy for the entire Ruhr region.

B. History of the Direktionsprinzip to 1766

The history of the mining law and the Direktionsprinzip, a system of state control over all mining activities, is intimately related to political developments in the Germanies. The Direktionsprinzip, to which much of German mining was subject from the seventeenth to the mid-nineteenth centuries, was the result of an historical evolution extending back to feudal law. The general privilege of mining, or the Bergregal, as it was already known in the twelfth century, was actually a composite right. Feudal law made a distinction between mineral rights and surface rights on the property on which the mine was located. The peculiar complexity of the medieval system was due to the fact that these rights were by no means always in the possession of the same person.[19]

The Holy Roman Emperor, from whom in feudal theory all rights were derived, had granted property rights to his territorial princes (Landesfürsten). Especially in the expanding eastern part of the empire, these rights were not necessarily accompanied by mineral rights. A strong emperor like Frederick Barbarossa still claimed the regalian right to all minerals as an attribute of his imperial sovereignty in the Roncagalia Constitution of 1150.[20] But unlike the French kings, who were able to keep the regalian rights intact because of a stronger central government,[21] the imperial government was too weak to resist the claims of its powerful princes. Consequently, the imperial regalia were carved up among them.

At first, the right to mine was granted formally, as a concession from the emperor, only to the most favored princes, but gradually more and more of them came into possession of the full privilege of the Bergregal. By the fourteenth century the emperor as a central power had become so weak that he could not even uphold his theoretical right to grant mining concessions. In 1356 Charles IV, in order to gain the allegiance and support of his territorial princes, forever sacrificed the imperial rights to mining. The Golden Bull of 1356 gave the regalian rights to all the Kurfürsten.[22] Finally, the Peace of Westphalia in 1648 recognized that all the Reichstände owned the regalian rights including mining rights.[23]

Yet the splintering of mining rights, which from the twelfth century are more properly called Hoheitsrechte,[24] or princely privileges, did not end here. Within the territories of the great princes the same splintering occurred on a smaller scale. Although technically these rights remained in the hands of the prince, in practice the tendency was for them to fall to the smaller estate owners. This splintering of regalian rights continued until some territorial princes managed to increase their power and consolidate it. This happened most significantly in Prussia. Here the prince was able to enforce the _Hoheitsrecht_--the actual exercise of the _Bergregal_. This meant state ownership of all minerals.[25]

In general the _Bergregal_ owners did not use the right themselves, but rather left the searching for minerals and actual mining to the private interest. The searching for minerals, the _Bergfreiheit_, had traditionally been free and left to interested individuals. But for the actual operation the mining interests had to pay dues, traditionally a tenth of the production of the mine. In addition the interested individuals received with the mining concession from the _Bergregal_ owner the right to mine on the field even against the opposition of the property owner. The property owner in this case had no rights over his own property.[26]

In general, during late medieval times, the influence of the _Bergregal_ owners on mining fluctuated. In times of economic prosperity, the influence was relatively small and limited itself to the traditional dues. During times of economic depression the influence increased the more the mining depended on the financial help of the _Bergregal_ owner. The state had always closely regulated the mining of precious metals and gradually over the centuries the state extended its influence to the mining of other valuable minerals, particularly coal. This increasing state involvement in the mining industry is clearly evidenced in the many mining codifications of the sixteenth century. For example, the Annaberger Mining Codification of 1509 forced private mine owners to contribute the necessary share of money to keep certain mines in operation even when no profit existed.[27]

"The so-called Direktionsprinzip was greatly strengthened after the Thirty Years' War when mining in many areas ceased to exist, and where it still survived it was in an extremely underdeveloped condition."[28] In the last instance it was also the result of the high dues demanded from an already weak mining sector. But there also was a lack of technical and business personnel. But perhaps more important, the growing absolutism favored the development of the Direktionsprinzip.[29] The ideas of mercantilism, whose basic principle was a planned intervention of the state into all economic activities, also affected the state's interest in the mining sector.

In the seventeenth century the newly acquired Prussian territories on the Rhine were under the Jülich-Cleve mining codification of April 27, 1542. On the basis of this codification, coal was made subject to the Bergregal.[30] Gradually the Brandenburg government took an interest in these territories and also in the existing coal mining industry.

Initially it appeared to be mostly interested in the dues and taxes from mining.[31] For this purpose it had appointed a mining official for this region. Dietrich von Diest, the second such official, appointed in 1632, gave a broad interpretation to the mining codification and tried to bring the Mark coal mining under stricter state regulation. He saw to it that the field concessions which up until then had been indiscriminately given were now granted in an officially prescribed manner, just as was done in ore mining. For a better enforcement, under his urging, the old codification of 1542 was republished on February 3, 1639.[32] But under the mining officials following Diest, the old habits reappeared. The mining officials again indiscriminately granted concessions and mines were badly worked, being alternately exploited and abandoned. Everyone mined as he pleased, no records were kept, and no one paid any attention to the mining codification.

Not until a hundred years later did the Prussian government show renewed interest in the County of Mark. This time it was not only the financial interest of the state but the general economic conditions that had changed. The nearby state-owned saltpeter works of _Königsborn_ at Unna were to be guaranteed a dependable supply of coal. In addition, the generally increasing wood scarcity and rising prices became of general concern.[33]

A government commission inspected the County of Mark in

1734, and the mining official Decker investigated coal mining in the Mark in the following years. He discovered that no one paid attention to the mining law. Decker tried to continue the work of Diest attempting to regulate coal mining. He demanded that records be kept and fields be measured before concessions were given to the mining companies. He set the number of shares (Kuxen) of a mining company at 128, and attempted unsuccessfully to gain an additional share--Freikuxe--for the state. His suggestion did provoke government intervention and led to a new codification of July 18, 1737, specifically intended for coal mining, and led to the opening in January, 1738, of the first Mark mining office at Wetter.[34] With this codification, the older tradition still practiced in parts of the Mark were to be suppressed and the first step was taken for individual mine management to fall under the control of a bureaucratic-fiscal mining administration. The costs to the state in opening a permanent mining office were transferred to the mining owner by means of special taxes.

The wars of Frederick the Great stopped all progress. In 1754/55 Minister Hagen toured the Mark and expressed the hopelessness of the mining conditions there. There were not enough technical personnel, trained miners and transportation facilities handicapping the sales of coal. Hagen observed that not even the wars, especially the Seven Years War, had stimulated private interests in coal mining. Consequently, the state revenue from mining was also low.[35]

To end these conditions, the Codification of 1737 was replaced by the Revised Mining Codification of April 29, 1766, for the duchies of Cleve and Meurs and the County of Mark.[36]

C. The Application of the *Direktionsprinzip* under Freiherr vom Stein

For most of the second half of the eighteenth century, the highest mining officials were also the highest political officials in the western territories. From the 1780's to the turn of the century, the Ruhr region was fortunate in having a most capable mining and political administrator in Freiherr vom Stein. He should be credited with the effective application of the *Direktionsprinzip* as prescribed in the 1766 mining codification. Under Stein, the *Direktionsprinzip* became an integrated and most important part of his general economic policy for the western Prussian territories and reached its fullest extent by the end of the eighteenth century.

Stein's service as a leading official in the western territories of Prussia was to last for twenty years. In 1780 Freiherr vom Stein was appointed to examine the navigational facilities of the Ruhr river. On February 16, 1784 at the age of 25 vom Stein was appointed Oberbergrath, the highest mining official for the western territories.[37] He held this position until 1798. During this period all the coal mines in the Ruhr region as well as Ruhr shipping were under his charge. In addition his other positions made him the highest political administrator. In 1793 he became President of the Mark Chamber, and in 1796 he was appointed Senior President of the chamber of all the western provinces of Prussia. This post he held until he became minister of commerce in October 1804.[38]

He was particularly interested in the development of the Mark coal industry and in the improvement of transportation. When Stein came to the Ruhr he had specific instructions from Minister Heinitz of May 20, 1784, ordering him to enforce the regulation of 1766, to stop the evasion in the payment of the coal tax, to check the wasteful exploitation of the coal, to discourage litigation between mine owners, to expand the market for the Ruhr coal, and to raise the standards of efficiency of the local mining administration.[39]

Stein himself believed that the basic objectives--to increase state revenue and to raise mining efficiency--could be accomplished by enforcing the Direktionsprinzip as delineated in the 1766 codification.

The two reports of July 26 and July 27, 1784, to Minister Heinitz indicate Stein's objectives.[40] The report of July 27, 1784, reflects Stein's effort to increase state revenues, not by increasing the taxes on the individual mine, but rather by improving direct state management of each mining operation. As the report indicates he wanted a complete revision of the mining office's practices in regard to the audit, accounting, and bookkeeping procedures. Further, his objective was direct state administered coal prices, wages, and other material factor prices, and to provide state financial help in the education of the miners. In the Report of July 26 he had already described ways to improve the state's mining administration, especially improving the state management of the individual mining operation.

Both reports were implemented by official instructions. In 1786 several of these instruction became effective. The mine owners were forced to hand over their bookkeeping records to the mining office. Further Stein enforced the state's administered coal prices. Finally the opening of a mining operation depended on the permission of the mining administration.

The biggest opposition by the mine owners was to the state mining administration's management of their enterprises. Many refused to comply despite the penalty of losing their right to operate the mine. The mine owners feared that all these reforms would lead to a rise in taxes. This opposition was not quieted until, with the advice of Stein, the state promised on November 29, 1786, that no further tax increases would occur.[41]

The other major objective was to raise the technical efficiency of mining, which Stein had argued would result only through specialization. Each mine should be supervised by a qualified foreman. Only skilled workers were to work at the coal face itself; all others were employed in other aspects of the operation. Stein found that the biggest obstacles to further development of the mining industry were the smallness of the mines, the lack of pumps, and the difficulty of transportation of the coal from the mines to the river Ruhr.[42] In 1784 he wrote to Minister Heinitz that the costs of coal mining were rising, that pumps were increasingly necessary, and that ignorance of machines was general. For this reason, Stein himself went to England in both 1786 and 1787, making inquiries regarding the use of steam engines to raise water. He was unsuccessful in introducing them. Not until 1801 was the first steam engine in the Ruhr mining area erected on the Zeche Vollmond. "Until late in the eighteenth century the coal was obtained from shallow open workings (Tagebau) or from drifts put into the hillside (Stollenbau)."[43]

Since Stein was also concerned with the sales of coal, the improvement of transportation became a major objective under his administration. The transportation development in the second half of the eighteenth century, the improvement of navigation of the rivers Ruhr and Rhine as well as the improvement of land transportation will be discussed in the third chapter. Let it suffice here to say that Stein was instrumental especially in the improvement of land routes. Under his urging two major highways linking the Ruhr with the rest of Prussia were built. Many secondary roads were also built, among them a vital road linking the two provinces of Mark and Cleve. In 1790 Stein personally succeeded in securing an agreement with the Abbess of Essen. Further, under the urging of Stein, who had earlier produced a plan for a coal depot at Ruhrort, the building of the large depot was begun in the 1790's. Ruhrort, at the juncture of the Ruhr and Rhine, had also come under state administration. The state now had an additional control over the sale of coal transported on the Ruhr River.

D. The Administration of the Direktionsprinzip, 1766-1851

The revised Cleve-Mark Mining Codification of 1766 introduced the Direktionsprinzip in the following manner:

> "Experience confirms that it is to the disadvantage of those interested in mining enterprises when the establishment and operation of a Gewerkschaft [mining company] is left to them alone; for the most part, mining operators must rely on inexperienced workers, mine supervisors and foremen who often lead them on to unnecessary operations. These operations frequently lead to losses of money, to misunderstanding between the Gewerkschaft and the workers, and to general disorder resulting in the bad reputation of our mining enterprises. Henceforth, all mining operations are to be conducted under the direction of the Bergamt [state mining office]."[44]

The Direktionsprinzip expressed a complicated interrelationship between the state power, state and private rights, and ownership of property. In general, rights are derived from power; in turn ownership is determined from rights. An essential characteristic of ownership is control of the property. If conditions exist where ownership is without the power to enforce the right to the property, then ownership has little meaning. Therefore, a basic question is not who owns the property, but who had effective control over the property.

The eighteenth century Prussian king had the power to enforce certain historically inherited rights, among them the mineral rights. Under the *Direktionsprinzip*, the state, in the person of the king, possessed the mineral right, *Bergregal*, and he effectively enforced it. But an effective enforcement also involved the control of the property where the coal appeared. In turn, to exercise the control over the mineral rights and property the establishment of an efficient bureaucratic mining administration was imperative. The property owner was left only with the right to demand compensations.

Consequently, although the king did not own the mining property, he effectively controlled it because he also possessed the power to enforce the mineral right. Under this system private ownership of mining property had only a very restricted meaning. The mining company, *Gewerkschaft*, is more properly described as sharing ownership with the state. The emphasis is not on the theoretical right that it was the sole owner of the mining property.

The purpose of this section is to outline the essentials

of the *Direktionsprinzip* as it was applied in the Revised Cleve-Mark Codification of 1766. The *Direktionsprinzip* was also clearly stressed in Part II of the *Allgemeine Preussische Landrecht*, in 1792. This part applied to those areas of Prussia where a specific mining codification did not exist. The emphasis will not be on the many and complex legal and technical details of this system of direct state administration, but rather on its major features. These include the bureaucratic administrative structure and its corresponding functions, as well as the general social and economic responsibility under the *Direktionsprinzip* for the Ruhr region.

The effective implementation of the general mining policy in Prussia involved an elaborate bureaucratic administration and a complex and detailed system of regulation, directing all aspects of the mining industry. In the Ruhr region after 1766, the effective implementation of the *Direktionsprinzip* and the orderly and efficient exploitation of the coal also depended on the establishment of a mining administration with capable and qualified officials.

The administration of the *Direktionsprinzip* as it applies to the Ruhr region was on five separate levels.[45] These ranged from the ministerial level to the management of the individual mining enterprise. In turn, the activities and responsibilities of each of these levels involved major areas of control. These ranged from the regulations concerning the acquisition of mining property to the responsibility of the mining administration for the improvement of transportation in the Prussian part of the Ruhr region. The latter was especially true until the end of the eighteenth century.

Even after 1766 the establishment of an effective and efficient administration was gradual. Although the codification provided the legal foundation and expressed a definite economic policy toward the Ruhr mining industry, it did not necessarily provide a clear basis for the establishment of an administrative apparatus, its functions and its jurisdiction. Interpretations, imaginative application, additional regulations and even new laws were necessary. In such a situation, certainly for the eighteenth century, a great deal depended on the leading mining officials and their understanding of the region's problems. However, the Ruhr was fortunate, at least until the end of the eighteenth century, in possessing such capable leaders.

Consequently, by the 1790's the basic structure and function of the administration had been worked out, although many jurisdictional problems remained. Further organizational changes occurred throughout the entire _Direktionsprinzip_ period, especially on the ministerial level, but in essence the basic structure and functions remained until the early 1850's.

After 1766 the general Prussian mining administration,

including the administration of the Direktionsprinzip, was in the hands of one of the king's ministers or cabinet officials. Also the king himself was not only theoretically but, as was frequently shown, actively the highest mining administrator. Below this level were the five mining headquarters, each representing a province or territory: Bonn, Dortmund, Breslau, Clausthal, and Berlin.

The provincial headquarters, Oberbergamt, for the entire Ruhr was at Dortmund. After 1803, beneath this office were two regional mining offices at Essen and Bochum. In turn each of the two regions was divided into districts. Finally, at the lowest level of administration, was the state management of each mining enterprise.

Before 1792, the highest mining administration for the Prussian part of the Ruhr region had been part of the general administration of the western territories, the Cleve Kriegs -und Domänenkammer. From 1792 to 1808, the mining administration for the Ruhr coal region became an independent subdepartment of the ministerial General-Oberfinanz-Kriegs-Domänendirectory at Berlin. Nevertheless, until about the end of the eighteenth century, the highest mining official also tended to be the highest political official in the Ruhr region. A great deal of freedom was left to this official, thus providing a unified economic and political leadership for the Ruhr region, never again to be achieved.

From 1808 to 1817, including the period of the French

occupation, the mining administration shifted from one department to another. From 1817 to 1834, the administration came under the supervision of the *Ministerium des Innern* whose major responsibility was trade and industry in Prussia. From 1834 to 1848, mining was under the jurisdiction of the Ministry of Finance (*Finanzministerium*). After 1848, it became part of the newly formed Ministry of Commerce (*Handelsministerium*).

These jurisdictional changes did reflect, if not a different policy, a different attitude toward the actual administration of the *Direktionsprinzip*. For instance, the *Ministerium des Innern*, whose major purpose was to encourage trade and industry, tended to have a more liberal attitude than the more conservatively inclined Ministry of Finance in the administration of the *Direktionsprinzip*. Certainly different interests were involved. These different attitudes are at least indirectly reflected in the mining reform attempts and transportation development discussed in subsequent chapters.

There were far fewer changes on the provincial and regional administrative level. Until 1803, there were a few changes in the location and number of mining offices in the Ruhr region. In 1792, the provincial headquarters at Dortmund was established. Previously, the mining office at Wetter had been the provincial as well as one of the regional headquarters. After Essen and Werden had been added in 1803, a regional mining office was established at Essen.

Politically, after 1815, the Essen-Werden region belonged to the Rhenish province, while the Mark mining region belonged to the Province of Westphalia, but the entire Ruhr mining region, including the offices at Essen and Bochum, were under the jurisdiction of the _Oberbergamt_ at Dortmund. This administrative structure existed until 1861, when the two regional mining offices fell victim to the mining reforms that had started in 1851.

In 1803 there was still no clear distinction between the administration function of the _Oberbergamt_ and the two regional offices. When the _Oberbergamt_ was established in 1792, its only official distinction was that it received instructions from the king or from the responsible ministerial office. But it also, simultaneously, carried out the regional function as one of the two mining offices in the Ruhr. When in 1803 the mining office at Essen was established, it became essential to clarify the various jurisdictions.

In April, 1803, the highest mining official in Prussia, _Oberberghauptmann_ von Reden, who had been previously the active official in Silesia, attempted to point out the most important activities of an _Oberbergamt_. Among them were the administration of the _Bergregal-Direktionsprinzip_, the handling of the field concessions, the administration of justice, i.e. enforcing the mining law and providing order.

Further, the *Oberbergamt* was to be concerned about the supply of a mining labor force and to safeguard the health and welfare of the miners. It was also responsible for the establishment and control of wages, coal prices, and taxes. It had to approve the operational plans for all the mines and controlled their bookkeeping records. Finally, it was directly responsible for the fast and efficient sale of the coal.

Basically, the *Oberbergamt* acted as the highest state authority in the Ruhr region, exercising supervisory and regulatory functions, overseeing the administration of the state mining law. But in the period from 1766 to 1803 it also had acquired the technical and business management of all the mines as part of its supervisional power. This created jurisdictional conflicts with the subordinate regional offices. Legally this was never resolved and caused continuous problems in the first half of the nineteenth century. In addition, the *Oberbergamt* did not possess the full powers of the state, but had to ask for ministerial directives and advice on the more important issues. Although the function of the *Oberbergamt* and the *Direktionsprinzip* in the Ruhr region were never clear, gradually after 1766 and especially after 1792, a tradition developed which often took the place of regulation.

Since the *Oberbergamt* had taken on most of the important functions, the two regional mining offices were left, after

1815, with the supervision and control of the mining companies and the accounting for the mines. They also regulated all employment problems, supervised the buying of the necessary material for the mines, as well as handling the sales of coal. They determined the profits to be distributed to the individual share-holders of a mining company and they also determined the Zubusse, that part which the shareholders had to contribute for the operating cost when no profit existed. Finally, they also supervised the measurement of the fields before a concession was granted. Of course some of the functions overlapped those of the Oberbergamt.

Each regional office, Bergamt, was divided into smaller districts. The head of a district was the Geschworene, who carried out the actual work delegated by the Oberbergamt or the regional office. Each such official was directly responsible for several mines, their operating plans, the size of the working force, all matters involving discipline and wages of workers. In general, this official supervised the management of the mines under his jurisdiction.[46]

The district official was assisted by two qualified specialists, the Obersteiger and the Oberschichtmeister. The Obersteiger was the technical expert. He was responsible for the actual mining operations. He gave the necessary instruction for the actual mining, set the actual wages, constantly inspected the mines and reported their conditions. Even contemporaries recognized that the welfare and order of the coal industry depended especially on the <u>Obersteiger</u>. He was the "soul of execution" for the entire <u>Direktionsprinzip</u>.[47]

The <u>Oberschichtmeister</u>, another specialist, was responsible for the non-technical aspect of mining. He supervised the business management, was responsible for the bookkeeping and general accounting. He kept a monthly journal for the mines under his supervision. Most importantly, he collected the taxes from the mines. This office existed until 1840.[48]

At the lowest level of administration was the actual mining operation of the individual mine and the privately owned mining company. In the service of the mining company were two officials, the <u>Steiger</u> and <u>Schichtmeister</u>. These two officials were responsible for the actual operation of the individual mine.[49]

The selection, employment and release of these officials was done by the regional mining office. But their salaries were paid by the mine owners for whose interest they were supposed to work. The <u>Steiger</u> was directly subordinated to the <u>Geschworene</u> or the <u>Obersteiger</u> and was the actual manager of the individual mine operation. The <u>Schichtmeister</u> was the business manager.

The <u>Steiger</u> directed, although with instruction from his superiors, the entire operation of the mine. He supervised the daily operation, the employment of workers, the upkeep of the buildings, the technical aspects of the actual mining. He was, consequently, the most important advisor to the district officials, since he was the one most knowledgable about the condition of the individual mine. Consequently, his advice was normally followed.[50]

The *Schichtmeister* was the "private manager of the mining property". He managed the business part of the operation, keeping the record of production, buying the necessary material for the mine, dealing with the buyers of coal, accounting for all buying and selling, paying the wages, etc.[51]

Under this system of administration private entrepreneurial activity was strictly regulated. The private mining company, *Gewerkschaft*, and its shareholders had few remaining rights. The *Gewerkschaft* had been for centuries the traditional organization for mining precious metals or other valuable minerals. Over the centuries, it had developed different characteristics and a complicated legal nature which in part maintained old features as well as adding new ones, adapted to new conditions. The *Gewerkschaft* was introduced in coal mining. The codifications of the seventeenth and eighteenth centuries, particularly in Prussia, made the *Gewerkschaft* the major business form and the focal point of state control over the mining industry. In the Ruhr coal mining region, the *Gewerkschaft*, its specific social, economic, and legal nature, was of course the product of the *Direktionsprinzip*.

Under the *Direktionsprinzip*, the company owned the mining operation but had little authority over it. It provided the capital and took the risk, but was explicitly by law not allowed to do anything on the property without the knowledge of the mining office. Although it paid the salaries of the *Schichtmeister* and the *Steiger*, it was specifically prohibited from interfering in the management of the mining operation. These two officials, however, were required by law to consult and listen to the advice or complaints of the mine owners. They especially had to consult the *Gewerkschaft* when greater costs or capital were involved in order to keep the mine in operation. Finally, each shareholder was legally obligated to pay money for the continuance of the operation, *Zubusse*, if no profits existed. If profits did exist, each shareholder received his part. At the same time each shareholder and the company as a whole was legally prohibited from interfering in the actual mining operation.

The share, *Kux*, in the *Gewerkshaft* was a concept from medieval times when the number of shares in a mining company had been equal to the number of persons participating in the mining operation. Historically the number of shares ranged from 12 shares in the twelfth century silver mining, to 128 shares by the eighteenth century. This number was also incorporated in the Cleve-Mark mining codification and later reaffirmed in the Allgemeine Preussische Landrecht. There were also additional shares, Freikuxen, which had to pay no Zubusse. One such free share was held by the church, the school system, and the miners organization, Knappschaft. These were transformed into money dues.

A share was legally treated as immobile property, literally being a physical share of the property. Legally, under the Direktionsprinzip, a shareholder could demand his precise physical share of the coal production. Although often not practical, this was possible until 1827 (for the Essen Bergamt region) when it was specifically prohibited. Further, the shareholder was not protected by limited liability, but rather was totally liable.

A shareholder could also sell his share or use it as a loan security, but it was legally a non-negotiable security in the stock market.

Decision making in the company was restricted. Unanimous approval by all shareholders was required by law for the more important decisions effecting the company. Such was the case when major costs were involved, when the mining operation was to be discontinued, or when the sale of the company was contemplated.

FOOTNOTES - CHAPTER I

1. Norman J. Pounds, "The Upper Silesian Industrial Region," Slavic and East European Series, XI (1958); Herman Fechner, "Geschichte des Schlesischen Berg-und Hüttenwesen in der Zeit Friedrichs des Grossen, Friedrich Wilhelms II und Friedrich Wilhelms III, 1741-1806," Z.f.B.H.S.-W. (1900, 1901, 1902); G. Felsch, Die Wirtschaftspolitik des Preussischen Staates bei der Gründung der Oberschlesischen Kohlen-und Eisenindustrie (1741-1871) (Berlin, 1919); William Henderson, The State and the Industrial Revolution in Prussia (1740-1870) (Liverpool 1958); R.v. Carnall, "Das Deukmal des Ministers Grafeu von der Reden bei Konigshütte," Z.f.B.H.S.-W., (1854), 201-224.

2. Felsch, pp. 1-20; Fechner, (1900), pp. 351-401; Pounds, pp. 16-17.

3. Pounds, pp. 22-37; Felsch, pp. 1-20.

4. Felsch, pp. 1-20.

5. Carnall, pp. 202-204.

6. Carnall, pp. 210-212; Pounds, pp. 35-37.

7. Felsch, pp. 1-4.

8. Pounds, p. 37, p. 71.

9. Henderson, p. 25.

10. But the large estate owners also exerted pressure and had received certain privileges under the 1769 Silesian Mining Codification. See Pounds, pp. 27-28.

11. Henderson, pp. 24-25.

12. Henderson, p. 25.

13. Henderson, p. 25.

14. Hans Spethmann, Das Ruhrgebiet (Berlin, 1933), I, p. 159.

15. Spethmann, pp. 158-59.

16. Spethmann, p. 158.

17. Spethmann, p. 157.

18. Spethmann, p. 160.

19. There has been much controversy as to the origin, development, legal interpretation, and application of the German medieval mining law. Some of the more important interpretations and legal histories are: Adolf Zycha, Das Recht des ältesten deutschen Bergbaues bis zum 13. Jahrhundert (Berlin, 1899); Adolph Arndt, Bergbau und Bergbau-Politik (Leipzig, 1894); Zur Geschichte und Theorie des Bergregals und der Bergbaufreiheit (Freiburg, 1916); Heinrich Achenbach, Das Französische Bergrecht (Bonn, 1896); Das gemeine deutsche Bergrecht in Verbindung mit dem preussischen Bergrecht (Bonn, 1871); C.J.B. Karsten, Über den Ursprung des Bergregals in Deutschland (Berlin, 1844); Ernst Schönbauer, "Beiträge zur Geschichte des Bergbaurechts," Münchener Beiträge zur Papyrusforschung und antiken Rechts-Geschichte in Deutschland (München, 1929); Hubert Ermisch, Urkundenbuch der Stadt Freiberg in Sachsen (Leipzig, 1883); J. Weiske, Der Bergbau und das Bergregal (Eisleben, 1845); R. Müller-Erzbach, Das Bergrecht Preussens (Stuttgart, 1916); Friedrich Luetge, Deutsche Sozial-und Wirtschafts geschichte (Berlin, 1952); Gerhardt Boldt, Staat und Bergbau (München & Berlin, 1950).

20. Achenbach, Deutsches Bergrecht, pp. 17-83; Arndt, Geschichte, p. 46; Schönbauer, p. 183; Karsten, p. 39.

21. Achenbach, Frauzösisches Bergrecht, for an early history and development of French mining law.

22. Johann F. Gmelin, Beiträge zur Geschichte des teutschen Bergbaues (Halle, 1783); Müller-Erzbach, p. 48; Weiske, p. 52; Boldt, pp. 1-2; Karsten, p. 44.

23. W. Westhoff-Schlüter, "Geschichte des deutschen Bergrechts," Sonderabdruck aus Z.f.Bergrecht, (1900, 45); Achenbach, Deutsches Bergrecht, p. 89.

24. Arndt, "Die Besteurung der Bergwerke," Z.f.Bergrecht, (1882, 67); Arndt, Bergbau, p. 173.

25. Herman Brassert, Berg-Ordnungen der Preussischen Lande (Köln, 1858). It contains the twelve mining codifications still in force in Prussia in 1850: Silesia 1769; Magdeburg 1772; Chursaxony 1589; Eisleben-Mansfelde 1763; Henneberg 1566; Jülich-Berg (Ruhr region) 1766;

Chur Cologne (Ruhr region) 1669; Nassau 1559; Nassau, so-called small codification, 1592; Hoburg Wittgenstein 1570; Chur-Trier 1564; Cleve-Meurs, Mark (Ruhr region) 1766; plus the sections of the Allgemeine Landrecht, 1790, dealing with mining, and the Code des Mines, 1810 of most of the Left side of the Rhine.

26. See also Pounds, pp. 16-17, for the mining law history in Silesia.

27. Boldt, Staats bergbau, pp. 2-4.

28. Boldt, p. 3.

29. Boldt, p. 4.

30. Th. Wagner, Corpus iuris metallici (Leipzig 1791). Contains the older mining codification including both the Jülich-Cleve mining codification of April 27, 1542, and the Annaberger Mining Codification of 1509. The Jülich-Cleve Codification used the Annaberger as a basis.

31. H.v. Achenbach, "Geschichte der Cleve-Märkischen Berggesetzgebung," Z.f.B.H.S.-W., (1869), p. 189.

32. Wagner, Corpus iuris metallici, p. 241.

33. M. Reuss, "Mitteilungen aus der Geschichte des Oberbergamtes zu Dortmund," Z.f.B.H.S.-W. (1892), p. 309.

34. The activities of Decker and his reports are contained in: Achenbach, "Geschichte...", Z.f.B.H.S.-W. (1869), pp. 193-199.

35. Reuss, "Mitteilungen...", pp. 310-311.

36. Brassert, Bergordnungen, pp. 107-147.

37. M. Lehmann, Freiherr vom Stein, Vol. I, p. 47.

38. G. H. Pertz, Aus Steins Leben, Vol. I, pp. 11-74.

39. Lehmann, p. 47; Henderson, p. 34.

40. Botzemhart, Eric, Freiherr vom Stein, p. 17; Reuss, "Mitteilungen...", p. 321.

41. Reuss, p. 321.

42. Henderson, p. 34.

43. Pounds & Parker, p. 49.

44. Cleve-Mark Mining Codification of 1766, Article XXIX, Paragraph 1; in Brassert, _Bergordnungen_.

45. Achenbach, "Geschichte...," _Z.f.B.H.S.-W._ (1869); M. Schulz-Briesen, _Der Preussische Staats-bergbau im Wandel der Zeiten_, Vol. I.

46. _1766 Codification_, Article XLIII, SS 1-6, Article XLIV, SS 1-10.

47. Achenbach, "Geschichte...," p. 218.

48. _Instruktion für die Oberschichtmeister, November 25, 1832_, O.B.A.-24/4.

49. _1766 Codification_, Article XLV, SS 1-2; _Allgemeines Preussisches Landrecht_, Part II, Article 16, SS 307.

50. _1766 Codification_, Article XLVI, SS 1-11, Article XLVII SS 1-14.

51. _1766 Codification_, Article XLV-XLVI, Article LII-LV. _Landrecht_, Part II, Article 16, SS 314.

CHAPTER II

OBJECTIVES AND REGULATIONS UNDER THE *DIREKTIONSPRINZIP*, 1766-1850

The last chapter discussed the mining administration and its functions. The purpose of this chapter is to analyze and describe summarily the operations of the *Direktionsprinzip*. An appendix to this chapter provides the basic statistical data for the Ruhr region's coal industry, 1766-1850.

The *Direktionsprinzip* as a state policy had several major goals. The principal goal was to develop the Ruhr mining industry. To achieve this the state encouraged a private entrepreneurial class and pooled resources with private enterprise. The state provided qualified personnel and private enterprise provided capital. Another goal, chiefly in the second half of the eighteenth century was the development of an integrated and independent regional economy based on the mining industry. Of course, the insurance of a high and constant state revenue through mining taxes was another major state goal and at times dominated the entire policy. Finally, the Prussian state was actively concerned with the welfare of the mine workers and the general public.

The implementation of these goals and the effective application of the mining law was carried out by the establishment of an elaborate and detailed administration. In turn this administration, in accordance with the state goals, had three specific objectives. The first objective was the control of the production of the industry. The second objective was to enforce the efficient and orderly mining operation for the individual mine as well as for the entire industry. The third objective was to provide a market for the individual firm as well as for the output of the industry.

This chapter will describe the achievement of these three objectives by the mining administration and their relation to the major policy goals of the state. A complete system of regulation was established, based on the legal provisions contained in the 1766 Cleve-Mark Mining Codification, the mining section of the Allgemeine Landrecht of 1792, and in supplementary regulations.[1]

A. Production Controls

The control of production had as its major purpose to adjust the output of the industry in accordance with apparent market demand for coal at any given time. The control of production involved the control of the number of firms and the size and scale of the individual operating firm in the industry.

In turn, the control of the number of firms in the industry involved the control of entry and exit of firms. The entry of a mine was the most strictly regulated aspect of the state's attempt to control the mining industry. This was because the mining administration had guaranteed that once a mine operated, its output would not be fixed except under extreme market conditions. Each operating firm was to be assured a cost-recovering and profitable income. Therefore, the strict regulations for entering the industry consisted of four basic steps, each requiring a legal permission. These were Schürfen, Muten, Verleihen, and Betriebserlaubniss.

Each step has a long and complicated legal history but each was finally defined in the mining codifications of 1766. Schürfen basically expressed the Bergfreiheit, the theoretical right of any individual to search for coal, although the preparatory work which was actually involved in proving that the coal in fact existed already required a legal permit (Schürfschein).[2] Furthermore, it described precisely the area--hills and valleys, etc.--where prospecting could take place. The permit, holding out hope for an eventual field concession, was valid for one and one half years but was also, under exceptional circumstances, renewable.[3] During this time no one else could search or claim a field in the area described by the permit.

It is significant that the state encouraged private

initiative at this level by indiscriminately allowing any interested party to search for minerals. Of course, if the state was interested in gaining an income through taxes from its mineral right, then it had to encourage the discovery and eventual exploration of minerals. However, even here, the state mining administration was interested in order. Therefore it precisely defined those areas where searching was allowed and did not give permits for large areas so as to discourage speculation with the fields.[4]

For the same reason, it was stated in the ministerial order of November 11, 1822, that the state would not give individuals a search permit for an additional field unless this field was a certain prescribed distance away from another. To strengthen this regulation further, neither were close relatives such as wives and children, brothers and sisters, given permits within the prescribed distance.[5]

Once the interested party had obtained a _Schürfschein_, he could apply for the concession of the field. The legal tradition gave the first finder the right to be the first one to claim the field.[6] The mining office now had to investigate whether the individual in fact had found coal. The claimant had to prove that his claim had indeed coal worthy of being worked. To do this the mining law required that he literally "uncover the coal to eye-sight."[7]

Again, in order to prevent speculation with the claims, the state demanded that this uncovering of the claim, _Muten_, be started within a period of four weeks. If this was not done, the claim was lost and the field again became free. The claimant was under the constant surveillance of the district officials. Only when exceptional difficulties arose in the uncovering work, could the time be extended by the Bergamt, and the Oberbergamt could give any number of extensions.[8]

In the area where the coal seams were near the surface or even cropped out, as was the case until the end of the eighteenth century, fulfilling the conditions for a claim was fairly easy. Later, as the coal had to be sought deeper under the ground and especially when the coal was under a water-bearing chalk layer, proving the existence of coal became more difficult and costly. The problems of the shaft mine and its relationship to the mining administration will be discussed in chapter IV.

During the first half of the nineteenth century, even the mining administration was aware of the problems connected with a claim and realized that not all claimants possessed the capital or could raise it in order to prove the existence of coal. The officials recognized the work, capital and risk was in no proportion to the possible gain by the private individual who was, at the most, given the right for a field concession and was in no way guaranteed the operation of the mine.[9] However the law was not changed until 1853 when the results of drilling were accepted.

Once the conditions for the claim had been met, the claimant had the right to demand a field concession which was either a 'long' or a 'square' field determined by the Bergamt. The fields were measured according to the legally prescribed size and shape. The size of the field had been appropriate for coal mining until the end of the eighteenth century.[10] However, when mining at the beginning of the nineteenth century involved greater depth, and the costs increased due to the water and ventilation problems, and new production methods, the old field size no longer was efficient.

The law of July 1, 1821 replaced the old regulation concerning field size in all the provincial mining codifications and the Allgemeine Landrecht section. It was intended to accommodate the new mining conditions. Under this law the new fields were to be square and much larger than the older ones. Unfortunately, the law was so unclear in its procedure for the actual measuring of the field that a dispute started between the Ministerium and the mining offices which finally led to a conference in 1834. The dispute was finally solved with the participation of the mining companies in 1836. Consequently, the law which was to benefit the new shaft mines could not be applied until 1836, and then was, in practice, not applied by the mining offices until the 1840's.[11]

The Betriebserlaubniss, the necessary legal permission

to begin the operation for the mine, provided the mining administration with a most effective tool to regulate directly the number of firms in the industry. This of course also effectively regulated the volume of production for the entire industry.

Already in 1783, a ministerial directive temporarily prohibited the _Bergamt_ from granting the _Betriebserlaubniss_ for new operation and operations of re-opened mines, with the explanation that the already operating mines had to be guaranteed a cost covering sale.[12] Again in 1803, when Essen and Werden fell to Prussia, the state reaffirmed that only it could grant the permission to start the operation of a mine and anyone disobeying would be penalized with a fine.[13] Once again, in 1821, the state emphasized in the instruction to the _Oberbergamt_, "not to permit a new mine to operate unless the existing mines could not fill the market demand for coal."[14]

Although, normally, anyone could acquire a field concession after having met the various prerequisites, the _Betriebserlaubniss_ was solely at the discretion of the state. Therefore, a mining company which may have invested a sizeable capital in the process of gaining a field concession and had made the necessary preparation for the operation did not have the right to start it. Chapter IV will contain a more detailed discussion of _Betriebserlaubniss_ during the first half of the nineteenth century when it increasingly became a hindrance to the larger shaft mining operations.

Not only did the mining administration successfully control the entry of the firms but it also attempted to regulate the exit of an operating firm. It tried to enforce the continuous operation of the individual mine, Betriebszwang, and it further enforced the payment of the legally required Zubusse from the mining company if the normal operation required it and if no profits existed.

Through the Bergregal, the state had appropriated the mineral rights from the property owner. Consequently, it felt responsible for assuring the proper exploitation of the minerals, so as to benefit the public and provide a state revenue. Therefore, it forced each mine to "constantly and continuously operate," the Betriebszwang.[15] Each mine had to have a minimum number of miners continuously at work and the mine had to be in operation a required number of days in the week. If the frequent inspections revealed that this was not the case, the mining operation involved was notified to correct this situation within four weeks under the threat of losing the field concession. There were also circumstances when mines were allowed not to operate because of water, technical, or sales problems, but in all cases this had to be approved by the mining office.[16]

This Betriebszwang forced the mine to stay in operation or forfeit its entire investment. The mining company was coerced into operating until either the cost of continued mining had become too much or the coal supply was exhausted. The latter, of course, was the administration's ideal goal.

To control the volume of output of the industry, it was not sufficient to control the number of operating firms in the industry. In addition, one of the goals of the mining administration was to have a constant and easily controlled flow of production, proportioned to the market demand, so as to guarantee every firm a market share. Consequently, the administration avoided, at all costs, extreme fluctuations in the output of the industry. In part, the control over entry to and exit from production of firms achieved this. But particularly in the early nineteenth century, new conditions in mining introduced the problem of the large output mine, the shaft mine. Consequently an additional objective was the control of the size and scale of the operating firms.

As already indicated above, consolidation of fields during the various steps in the acquisition of a concession was strictly regulated. The consolidation of firms already operating involved legal and practical problems.

Since the size of mining fields, especially in the nineteenth century, proved to be too small for shaft mining, consolidation of fields was a tempting goal. However, the law did not provide for consolidations; at least no provisions were in the 1766 codification nor in the _Allgemeine Landrecht_. Although the mining administration in principle did not approve consolidation, particularly not for mines producing profitably, in practice it allowed and even favored consolidations if conditions required it. This was the case when the continuous operation of a mine was involved, as was true of many small mines. In this case, the biggest obstacle toward consolidation of neighboring fields was the legal nature of the mining company itself. It required the unanimous approval of all the share-holders.[17] The initiative for consolidation could come from the company but all other aspects were handled by the mining administration. Occasionally the initiative would also come from the *Bergamt* itself. In that case it tried to persuade the company of the benefits resulting from consolidating their neighboring fields.

In 1840 there was even a reduction of certain taxes when consolidation took place. The consolidated mine would not have to pay taxes for each mine separately but only as one single mine. Nevertheless, the major obstacle in consolidation was not necessarily the mining office but the unanimity requirement. This law was changed in 1851.

B. Managerial and Administrative Controls

The second major objective of the mining administration, in keeping with the state's policy goals, was to enforce the efficient and orderly operation of each mine as well as that of the industry. In order to achieve this end, especially in the eighteenth century, it supplied the necessary qualified personnel, provided the technical knowledge, and insured a supply of labor--the miners.

The qualified state personnel supervised the technical and business operation of the mines.[18] It directly supplied the technical and business manager, the *Steiger* and *Schichtmeister*, for the individual firm. Before a mine began its operation, the mining office directed the responsible technical official, the *Geschworene* or *Obersteiger*, to provide a plan of operation, *Betriebsplan*.[19] This plan contained a detailed description of the location of the mine in relation to the transportation routes and the nearest market area, recommended the procedure for working the coal, took into account the water problems and insisted on the proper ventilation. Finally, the *Betriebsplan* estimated the expected output, costs, the time before production would begin, and the profitability of the mine. Once a mine was in operation it was regularly inspected and once a year a general inspection by the head of the *Bergamt* or even of the *Oberbergamt* was undertaken. On the basis of these inspections, a *Betriebsplan* for the following year was drawn up and approved by the mining office. Only on the basis of such a plan was the actual operation of the mine allowed. The district official was responsible for the enforcement of this operation plan. The *Steiger* of the individual mine was to carry it out. If the plan of operation required new shafts, tunnels, the buying of the new equipment, etc., all involving increased cost to the mining company it had to pay. The company could complain and sometimes influence changes in the technical plan if the cost were too great.[20] However, normally the company could not change the operation plan or interfere in the actual operation of the mine itself; it simply was obliged to pay the required Zubusse.

The mining administration also was interested in reducing the cost of operation. Wood was an essential mining material, it was used for shoring up and for construction of underground coal ways. But by the early nineteenth century wood had become one of the most costly items in mining. In the ten years, between 1831 and 1841 the price of mining wood doubled. Consequently, the mining administration attempted to reduce the cost by experimenting with chemically treated wood which would last longer, using other materials such as iron or stones where possible, and particularly re-using wood from closed mines.

Further the mining administration tried to improve the underground operation by introducing iron rails. These, however, cost more than most mining companies were willing to pay; consequently, they were initially reluctant. For this reason, the Essen mining office itself had built for exhibition purposes its own underground iron rail system, complete with carts. Gradually, in the course of the first half of the nineteenth century, mines began to use iron rails.[21]

The establishment of a trained and experienced labor force, the basis for any efficient and orderly operation, was another major objective of the mining administration. During the second half of the eighteenth century, the state had successfully created a permanent core of professional mine workers. Historically, the miners had always had special privileges; this was also true under the *Direktionsprinzip*.[22] The ministerial instruction of May 16, 1767, listed these special privileges, which included the exemption from military service and exemption from all city and village obligations. The miners also came under the legal jurisdiction of the mining office. The mining administration's responsibility for the welfare of the miner covered all aspects of his life.[23]

Under the *Direktionsprinzip* the miners were under the direct supervision of the regional offices.[24] These regulated and supervised the employment, including the hiring, changes in the place of work and dismissal of workers. Before a miner was hired he had to sign an official registration record. This entitled him to belong to the miners' fund. This fund had been established in 1767 and was supervised by the state. It was basically a health and accident insurance as well as a pension fund. Both the mining company and the miners contributed a fixed share.[25]

Once a miner was properly registered he was bound to the contract which he had made with the *Schichtmeister* of the individual mine. He could not easily quit or change jobs.[26] Furthermore he had to work in any mine the mining office placed him. His place of employment was at the discretion of the district officials,[27] who were instructed to adjust the employment at the mines under their supervision according to market conditions. This was done infrequently and then only when individual mines faced depressed markets. The objective was not to fix the output of the individual firm but rather to assure the continuous operation of the mines in a given area under adverse market conditions.

The mining office classified the miners according to age and experience, defined their jobs and established the wages of the individual miner accordingly. Consequently, under the regulations of the _Direktionsprinzip_, the miners were practically independent of the private company for which they worked and which paid their wages. The miners regarded the state rather than the private mining company as their employer. The mining office had hired them and as a condition for employment had even demanded from them an oath of loyalty to the state.[28] Although the miners under the _Direktionsprinzip_ were protected by the state, they also had lost certain basic personal freedoms. They did not have the right to choose or change jobs, nor could they easily leave the mining profession. These freedoms they gained only in 1860.

C. Market Controls

The last major objective of the mining administration was to provide a market for each operating firm and for the output of the entire industry. Furthermore, in the second half of the eighteenth century it felt particularly responsible for expanding the market for the increasing output of the industry.

The economic experience of the eighteenth century appeared to indicate that in the long run the industry's market could not be greatly expanded. Although the market could be expanded in the short run, the major obstacle was an inadequate transportation system. Consequently, especially in the second half of the eighteenth century, the mining administration had a separate but closely related objective: the improvement of transportation in the Ruhr region. The general transportation development will be discussed in Chapter III.

Regulating the sale of coal had two major purposes: to guarantee a market share for each operating firm, and to ensure relatively constant state tax receipts from the production of coal. The eighteenth century coal market was highly fragmented. Inadequate transportation and communication resulted in large price variation from one locality to another. Further, the demand in a particular local market was fairly constant and predictable. The typical situation in the eighteenth century was that a small number of mines in a given locality, using the same transportation routes, supplied the same market. Since the number of operating mines in such a locality had already been restricted by the administration's control of entry, it was now responsible for ensuring each mine an adequate return by providing an appropriate share of the local market. To achieve this, a system of tax and price regulations was instituted.

The basic guide for the price and tax policy was the law of October 26, 1755, which was repeatedly reaffirmed by instructions until the turn of the century. It was still in force in Westphalia in 1850. According to this law,

> "the _Bergamt_ should revise the taxes and prices of the coal from all mines in such a manner as to take into account the quality of the coal, the geographic location of the mine, the ability of a mine to make profit, and the access of it to the market."[29]

Under this law, the mining office's function was to set prices and taxes in such a manner that competitive mines could operate next to each other. Specifically, differential prices and taxes for mines in the same locality were to cancel out any "natural" advantage or disadvantage, especially due to location in relation to the market.[30] In general, the basic philosophy underlying the administration's price and tax policy regarded competition as harmful and price competition as most harmful. Low prices were regarded as disastrous because they were expected to drive out other mines. High prices were harmful to the consumer. The administration felt responsible for safeguarding the welfare of the consumer as well as that of the industry.

Although the *Bergamt* was ultimately responsible for the establishment of the coal prices, in practice, until the early nineteenth century, the actual price determination was handled by the district officials.[31] These officials, supervising a relatively small number of mines, had the best view of the local production, sales, and market conditions. These officials, with the help of the *Steiger* and *Schichtmeister* of the individual firm, were able to judge the advantages and disadvantages of each mining operation. They took into account the location of the mines, transportation routes, the direction of sales, and the market conditions for each of the mines, and determined the prices of coal accordingly. Therefore, each district tended to have its own general price level. Given these conditions, the regional mining offices and the *Oberbergamt* were also quite capable of viewing the general production, sales, market, and transportation conditions for the entire industry and could act as general price regulators.

A specific price of coal was determined for each mine. The quality of the coal, a normal profit, and the cost of production, which included among other factors the geological condition and ease of working the coal, determined the price of coal for each mine. The largest variation in the price between mines was caused by the location of the mine in relation to the nearest transportation route, the distance to the market, and the type of market. Furthermore, one mine could have several prices. Even the same seam could have different prices because the sale of the coal involved different markets, transportation routes, or different means of transportation such as water or land transport.

Part of the price policy objective was the establishment of a mine price that would at least cover the cost of operation during a given year.[32] The cost of production normally included the yearly cost of operation, but it did not necessarily include the initial outlay of capital. However, during the eighteenth century, the investment necessary to start an operation was relatively small. Stollenbau, putting drifts or tunnels into the side of a valley or hill, and Tagebau, open pit workings, frequently only required just the basic tools such as pick and shovel and a small building. Therefore, the normal profit tended to include the repayment of debts made in starting an operation or even to continue operation.

However, the administration's policy did have as an objective the inclusion of a reasonable profit as a determining factor in the price of coal for each mine. In the short run, this profit was to provide a sufficient incentive for the private mine owner or company to stay in operation. In the long run, the profit was calculated to maintain and encourage the incentive, initiative, willingness to take risks and to attract investment from the private entrepreneur in the mining industry. The state definitely wanted an active private entrepreneurial class in the industry to supply the capital.

The other factor determining the price of coal was its quality. For most of the Direktionsprinzip period, the coal was only vaguely and unscientifically classified. There were only two general categories. Quality coal, Melirte Kohle, could be exported. It could endure long distance land or water transport without becoming powder. This coal was usually in larger pieces. Gruss or Schuttkohle was powdery, small pieced and generally inferior in quality.

The external factors which were taken into account when establishing the price of coal for a mine involved transportation and specifically the ease and cost of transportation to the market. Again, the basic principle of the price policy was to account for the difference in location and, thereby, the difference in transportation cost for each mine in such a manner that the disadvantaged mine could compete on the market. A higher or lower price at the mine was to cancel out the difference in transportation cost between mines.

Therefore, under the Direktionsprinzip the state established price at each mine was regarded by the administration both as the actual sales and market price of the coal.

Until the early nineteenth century the actual sale of coal tended to take place at mine site. The coal was actually bought with cash at each mine. The typical situation in a given locality was that the buyers would provide their own transport or pay for the transportation. The local consumer, such as the blacksmith, household user of coal, or a local small industry normally transported its own coal. The consumer in the vicinity of several mines was perfectly free to choose the mine from which he would buy his coal. The coal for more distant markets was handled by transporters using horses as draft or pack animals or it was shipped on the Ruhr River.

Since the roads were extremely bad in the vicinity of most mines and only a few major roads were in good condition, long distance land transport was difficult and expensive. For instance, even by 1835, the price of coal in the _Hardenbergsche_ district, south of the Ruhr River, was 35 pfenning per _Ztr_. at the mine, but at Elberfeld only 30 km away, the price was 105 pfennig. Most of the difference was transportation cost.[33]

Finally, the tax policy was also closely related to the price policy. Under the _Direktionsprinzip_ in the Ruhr region the _tenth_ was the largest and constant share of all taxes. The taxes and chiefly the _tenth_, had a very special historical and legal development. From the historical point of view, the _tenth_ represented a share in the production rather than a simple tax.[34] In medieval times, the property owner and frequently the owner of the mineral rights such as a territorial prince participated directly in the mining of precious metals and valuable minerals. The later mining codification of the seventeenth and eighteenth centuries maintained some vestiges of this earlier practice. For instance, in all the Prussian mining codifications except for the Cleve-Mark Mining Codification of 1766, the property owner had the right to participate up to one half in the mining operation.[35] This was the case in the Silesian Mining Codification of 1779.[36] In the eastern provinces the large estate owners had the political power and had received legal concessions from the king.[37] Further the state maintained the right to exploit minerals on its own account. In Silesia, mines and iron works were owned by the state.

In the western parts of Prussia, in the Rhineland, large estate owners were few and the aristocracy not powerful nor was the Prussian state willing to engage in mining on a general scale. Of course even in the western parts, the state engaged in mining. In the Saar, the state owned the coal mines and even in the Ruhr region there were state owned mining enterprises such as the saltpeter works near Unna. Nevertheless, the state limited its ownership of the minerals to the controls under the _Direktionsprinzip_ and its claim to only one-tenth of the mining production. Here the state demanded the _tenth_ on the grounds that it owned the minerals and had leased them to private interests. The property owner had no legal claim to participate but had to be compensated.

Under the Direktionsprinzip in the Ruhr region all taxes were based on the established price at the mine. The tenth was one-tenth of the value of the output but not the actual sales value of a mine. This crucial difference made the tenth an output stabilizer guarding against overproduction during a given year. Consequently the tenth did contribute to the administration's objectives of a smooth flow of production and constant state revenue.

All taxes were either paid into the state mining treasury, the Zehntkasse, or paid into the mining administration fund, the Bergamtskasse.[38] The Zehntkasse was the actual state income. It consisted of the tenth and the Freikuxgeld. Traditionally the state or the territorial authority also received two free shares in the mining company. These shares were in addition to the normal number and did not pay Zubusse. With the 1766 codification these shares were converted into the equivalent money value and became a constant percentage of the coal sold by each mine.

The Bergamtskasse provided the money to operate the mining administration.[39] The Quatembergeld, specifically, provided the money "to pay the salaries of the mining officials and to provide the necessary money for the upkeep of the mining office." A constant part of the Quatembergeld, which was a constant percentage of the value of production of each mine, was the Messgeld. This was a levy which was paid by all mines for the measuring of a field before a concession had been granted. The Rezessgeld was a tax paid by all mines, both operating and non-operational. It was basically a property tax. The regular quarterly payment by the private mine owners was meant to indicate that the field was still in their possession and, in the case of non-operational mines, that there still existed the desire to keep the concession. The major portion of the Gedingegeld were fees paid at the various stages before gaining a field concession. In general, for the Ruhr region, the total taxes under the Direktionsprinzip amounted to about 15 per cent of the average value of output per mine from which the state income was approximately 11 per cent.

The price and tax policy objective was to regulate the sale of coal and the income of the mines. The mining administration accomplished this as long as it effectively controlled the sale of coal from each mine and it was capable of determining the different prices of coal for the mines. For most of the Direktionsprinzip period the administration was successful in controlling the sale of coal by each mine, but it did not necessarily control the resale of coal after it had been sold at a mine.

By the 1820's the administration was confronted with increasing difficulty in enforcing its price and tax policy objectives. It was losing control over the sale of coal on the market through the re-sale by mine owners and coal dealers. Further, the establishment of prices became extremely difficult if not impossible in an improved and expanding market. In general the elaborate administrative structure and its complex regulations began to break down in the course of the first half of the nineteenth century. The problems confronting the mining administration will be dealt with in the next chapter.

APPENDIX TO CHAPTER TWO

Tables 2.0, 2.1, and 2.2 provide the general statistical information for the entire period 1766-1850.[40] These data are based on the official Oberbergamt Dortmund statistics. Table 2.0 contains only production data, since no other sources are available until 1788.

The sudden jump indicated in Table 2.1 between 1803 and 1804 was due to the political acquisition by Prussia of the Essen-Werden region. The mines in this area also came under the *Direktionsprinzip*.

TABLE 2.0

PRODUCTION OF COAL: 1764-1787

Year	Production (1000 tons)	Year	Production (1000 tons)
1764	52	1776	87
1765	56	1777	83
1766	56	1788	88
1767	63	1779	95
1768	58	1780	100
1769	73	1781	102
1770	70	1782	119
1771	63	1783	123
1772	71	1784	111
1773	69	1785	124
1774	70	1786	111
1775	80	1787	124

Source: M. Reuss, "Mitteilungen aus der Geschichte des königlichen Oberbergamt Dortmund," _Festschrift_, p. 14.

TABLE 2.1

PRODUCTION, MINES, EMPLOYMENT, VALUE OF SALES: 1788-1850

Year	Number of Mines	Total Employment (100)	Total Production (1000 tons)	Total Value Sales (1000 R.M.)
1788	170	12	128	-
1789	184	12	132	-
1790	167	12	137	-
1791	158	12	132	-
1792	154	14	177	584
1793	159	15	181	600
1794	157	13	160	608
1795	146	14	162	581
1796	154	15	196	647
1797	150	15	201	757
1798	154	17	210	780
1799	149	17	228	867
1800	158	15	231	922
1801	158	13	201	871
1802	160	14	209	992
1803	140	16	201	871
1804*	229	31	380	1572
1805	195	31	392	1621
1806	206	28	353	1494
1807	229	31	338	1431
1808	195	29	339	1482
1809	183	29	351	1555

Table 2.1 - continued

Year	Number of Mines	Total Employment (100)	Total Production (1000 tons)	Total Value Sales (1000 R.M.)
1810	177	31	369	1573
1811	186	31	343	1562
1812	189	32	344	1639
1813	178	29	337	1512
1814	174	32	333	1551
1815	173	31	388	1766
1816	175	34	427	1927
1817	168	34	434	2056
1818	167	33	424	2048
1819	166	33	404	2122
1820	161	36	425	2102
1821	145	34	408	2115
1822	143	36	495	1956
1823	144	36	393	2035
1824	156	37	408	2192
1825	166	39	437	2381
1826	161	40	455	2412
1827	168	41	468	2654
1828	172	43	499	2685
1829	165	42	522	3060

Table 2.1 - continued

Year	Number of Mines	Total Employment (100)	Total Production (1000 tons)	Total Value Sales (1000 R.M.)
1830	172	45	511	3145
1831	180	47	626	3520
1832	194	52	675	3218
1833	211	54	761	4342
1834	216	54	767	4274
1835	190	59	784	4371
1836	199	68	745	3918
1837	208	75	868	4540
1838	209	81	996	5694
1839	213	83	1007	5871
1840	221	90	990	6006
1841	213	92	1092	6671
1842	226	94	1130	6317
1843	211	94	1080	6461
1844	214	101	1195	7134
1845	215	105	1265	7523
1846	219	109	1345	7747
1847	226	119	1440	8614
1848	221	115	1337	7714
1849	207	121	1384	8093
1850	198	127	1666	10000

Source: Reuss, "Mitteilungen...", _Festschrift_, pp. 14, 83.

TABLE 2.2

1788-1850 AVERAGES: PRODUCTION PER MINE, EMPLOYMENT PER MINE, PRODUCTION PER WORKER, AND PRICE

Year	Average Production Per Mine (tons)	Average Production Per Worker (tons)	Average Employment Per Mine	Average Price Per Ton (R.M.)
1788	753	107	7	-
1789	717	110	7	-
1790	820	114	7	-
1791	1468	110	8	-
1792	1149	126	9	3.63
1793	1138	121	9	3.97
1794	1019	123	8	4.05
1795	1110	116	10	3.97
1796	1273	131	10	4.04
1797	1340	134	10	4.02
1798	1364	123	11	4.26
1799	1530	134	11	4.49
1800	1461	154	9	4.51
1801	1272	155	8	8.06
1802	1306	149	9	5.22
1803	1436	127	11	4.97
1804	1659	123	14	4.36
1805	2010	126	16	4.42
1806	1714	126	14	4.45
1807	1476	109	14	4.58

Table 2.2 - continued

Year	Average Production Per Mine (tons)	Average Production Per Worker (tons)	Average Employment Per Mine	Average Price Per Ton (R.M.)
1808	1738	117	15	3.78
1809	1918	121	16	4.82
1810	2085	119	18	4.62
1811	1844	111	17	4.94
1812	1820	108	17	4.96
1813	1893	116	16	4.84
1814	1914	104	18	5.17
1815	2243	125	18	4.86
1816	2440	126	19	4.92
1817	2583	128	20	4.95
1818	2593	128	20	5.12
1819	2434	122	20	5.68
1820	2640	118	22	5.36
1821	2757	120	23	5.36
1822	3462	138	25	5.58
1823	2729	109	25	5.41
1824	2615	110	24	5.43
1825	2633	112	23	5.77
1826	2826	114	25	5.66
1827	2786	114	24	5.85
1828	2910	116	25	5.73
1829	3164	124	25	6.21

Table 2.2 - continued

Year	Average Production Per Mine (tons)	Average Production Per Worker (tons)	Average Employment Per Mine	Average Price Per Ton (R.M.)
1830	2971	114	26	5.72
1831	3478	133	26	5.85
1832	3479	130	27	5.81
1833	3607	141	26	5.95
1834	3551	142	25	6.03
1835	4126	133	31	6.07
1836	3744	110	34	5.72
1837	4173	116	36	5.62
1838	4766	123	39	6.28
1839	4728	121	39	6.61
1840	4480	110	41	6.27
1841	5127	119	43	6.64
1842	5000	120	42	6.86
1843	5118	115	45	6.80
1844	5584	118	47	6.49
1845	5884	120	49	6.38
1846	6142	123	50	6.24
1847	6372	121	53	6.46
1848	6050	116	52	6.05
1849	6686	114	58	6.12
1850	8414	131	64	5.93

Source: Derived from Table 2.1.

FOOTNOTES - CHAPTER II

1. Brassert, Bergordnungen; contains 1766 Codification and Allgemeines Preussisches Landrecht (A.P.L.).

2. 1766 Codification, Article 1, SS 1; A.P.L., Part II, Article 16, SS 141.

3. 1766 Codification, Article 1, SS 2; A.P.L., Part II, Article 16, SS 144.

4. 1766 Codification, Article 1, SS 3; A.P.L., Part II, Article 16, SS 143.

5. OBA-68/nn (Münster, Archive).

6. Boldt, Staatsbergbau, p. 5; C.J.B. Karsten, Grundriss Bergrechts lehre, p. 89.

7. This was one of the most crucial requirements of the mining law for the later shaft mining operation. But most of the Prussian codification contained such a requirement. 1766, Article 3, SS 1, 2; A.P.L., Part II, Article 16, SS 162-168.

8. 1766, Article 6, S 2.

9. Krampe, p. 34.

10. 1766, Article 1, Article 9, S 2a; A.P.L. Part II, Article 1, S 156.

11. K. Oberste-Brink, "Die Vierung und ewige Teufe der Langenfelder nach dem preussischen Gesetz vom 1. Juli 1821, Glückauf, (1941), p. 649-51. Contains law of 1821. Boldt, p. 6. Brassert, "Die Materialien des Gesetzes über die Verleihung des Bergwerkeigenthums vom 1. Juli 1821," Z.f.B.H.S.-W. (1861), pp. 493-506.

12. "Instruktionen für das Cleve-Meurs-& Märkische Bergamt zu Wetter, 24. Mai 1783," in Reuss, "Mitteilungen...," Z.f.B.H.S.-W. (1892), pp. 317-318.

13. Patent wegen Verwaltung des Bergregals in...Essen-Werden, 12. April 1803, O.B.A.-522, Münster Archive.

14. Oberberghauptmannschaftliches Rescript an das Oberbergamt 29. Sept. 1821, O.B.A.-68/3, Münster Archive.

15. 1766, Article 7, SS 2; A.P.L., Part II, Article 16, SS 193, 194.

16. 1766, Article 7, SS 1; A.P.L., Part II, Article 16, SS 201, 202.

17. Klostermann, "Bemerkungen über die Mobilisierung der Kuxe," Z.f.B.H.S.-W. (1861), p. 315.

18. 1766, Article 29, SS 2, specifically stated that the mining office would direct all mining operations: "...Unter des Berg-Amtes Direction alle Zechen betriebeu werden sollen."

19. 1766, Article 29, SS 1-2; Article 32, SS 1-3; A.P.L., Part II, Article 16, SS 82, 83.

20. Landrecht, Part II, Article 16, SS 82.

21. Krampe, Staatseinfluss, pp. 59-66.

22. Wagner, Corpus iuris metallici, pp. 1263-1267; contains: "General privilegien für die Bergleute im Herzogtum Cleve, Fürstentum Meurs und der Grafschaft Mark vom 16. Mai 1767," and "Instruktion zur Errichtung und Führung der Knappschaftskasse für Bergleute im Herzogtum Cleve, Fürsteutum Meurs und Grafschaft Mark vom 16. Mai 1767."

23. Zycha, "Ein altes soziales Arbeiterrecht Deutschlands," Z.f. Bergrecht, (1900), p. 455.

24. 1766, Article 48, SS 1, 8.

25. See note 22.

26. 1766, Article 44, SS 8.

27. Reglement für Bergleute, 30. April 1781, SS 1, 9, Mining Library Essen: 8394a2, Photocopy.

28. Krampe, p. 130.

29. Carnall, Die Besteurung der Bergwerke in Preussen, p. 50.

30. Reuss, "Mitteilungen," contains "Instruktion für das Cleve-, Meurs-und Märkische Bergamt zu Wetter vom 24. J Mai 1783", Z.f.B.H.S.-W. (1892), p. 319.

31. 1766, Article 73, SS 2.

32. Carnall, p. 50; Klostermann, Allgemeines Berggesetz, p. 48.

33. Spethmann, Ruhrgebiet, p. 221.

34. Arndt, "Die Besteurungen d. Bergwerke," Z.f. Bergrecht (1882).

35. W.J. Brassert, "Mitbaurecht bis zur Hälfte," Z.f.B.H.S.-W., vol. IV, pp. 1-28.

36. Pounds, Upper Silesia, p. 26.

37. Pounds, pp. 26-28.

38. Carnall, Die Bergrecht in Preussen und deren Besteurungen (Berlin, 1850); Entwicklungen des Nordrhein-Westfälischen Steinkohlenbergbaues in der zweiten Hälfte des 19. Jahrhunderts, vol. XII, pp. 266-270.

39. Entwicklungen, vol. XII, pp. 267-268.

40. The basic statistical sources are from:

Reuss, "Mitteilungen aus der Geschichte des Königlichen Oberbergamtes zu Dortmund und des niederrheinisch-westfälischen Bergbaues," Festschrift zur Feier des 100-jährigen Bestehens des königlichen Oberbergamtes zu Dortmund, 25. Juni 1892, (Berlin 1892).

Die Entwicklung des niederrheinisch-westfälischen steinkohlenbergbaues in der zweiten Hälfte des 19. Jahrhunderts, ed. Verein für bergbauliche Interessen im Oberbergamtsbezirk Dortmund, 12 vols. The economic development, information and dates are contained in vols. X, XI, XII.

Reuss, "Mitteilungen," is solely based on official OBA Dortmund statistics, and was written in behalf of the Oberbergamt. All production, employment, number of mines, price, sales statistics are taken from this work, pp. 14, 82, 83.

CHAPTER III

THE STATE AND TRANSPORTATION IN THE RUHR

A. Introduction

Under the _Direktionsprinzip_ one of the major objectives of the mining administration was the provision of a market for each operating firm and for the output of the entire industry. The state as well as its agencies became directly involved in the development of transportation. After all, transportation cost is a major element in the cost of coal to the final consumer. Furthermore the expansion of the coal market in the long run depended on improvement in transportation.

The role and attitude of the state and its agencies toward transportation and the transportation development in the Ruhr region itself are the major topics in this chapter. The attitude and policy of the state toward transportation underwent dramatic changes from the eighteenth century to the first half of the nineteenth century. In the first period the state had assumed a dynamic and leading role. The _Direktionsprinzip_ and transportation were part of a general coordinated economic policy for the entire Ruhr region.

In the first half of the nineteenth century the state

tended to play a passive role, sometimes even hindering the improvement in transportation. In addition the state no longer gave the Ruhr region's transportation development a favored treatment, as had been the case in the eighteenth century. Furthermore the *Direktionsprinzip* was reduced to a mining policy and separate from transportation.

On the provincial level after 1815 the president of Westphalia, Ludwig von Vincke, did focus his personal energy on the improvement of transportation. Reflecting an earlier state attitude, he felt transportation was the best way to promote industrial development. But he soon discovered that his own enthusiasm for new highways was not necessarily shared by the government departments in Berlin nor by the local authorities in Westphalia. At least until the 1830's the state claimed it did not have the financial resources and that it had been forced to get loans for its own household finances. In the 1830's the financial situation in Prussia had improved and von Vincke was able to push ahead with some badly needed road construction.[1]

The mining administration in the Ruhr region had itself an ambiguous policy toward transportation. On the one hand, it favored improvement to guarantee and even expand the sales of coal for the entire industry; on the other hand, it was dedicated to the *status quo* of the *Direktionsprinzip*. The latter did not favor competition and the loss of control of production and sales through a constantly expanding market.

For instance the conservative leadership did not favor new modes of transportation such as the railroads. Finally the mining administration did not have any legal authority over transportation, but even the influence it did have, lacked coordination and effectiveness because of dissent in its own ranks and conflicts of interests between mining offices.

During the nineteenth century the government at Berlin, the provincial governments, and the Ruhr mining administration although all representing the state, frequently were in conflict. They had different interests and used different means to achieve different objectives.

In the first half of the century private interests were the major force for the improvement and development of transportation. The major burden in improving transportation in the Rhenish and Westphalian provinces and the Ruhr mining region, in the absence of effective state leadership, fell on the shoulders of progressive, farsighted, and profit seeking entrepreneurs. Gradually private enterprise was able to force the state to adopt a new attitude toward transportation, primarily toward the railroads. During the 1840's the state began to be actively involved in railroad construction.

In both the eighteenth and nineteenth centuries transportation development was directly related to the development of the Ruhr coal industry. As long as mining had taken place near the Ruhr River, river improvements remained the major object.[2] As mining moved northward and the large-output producing shaft mines came into existence, new and additional means of transportation became necessary. The Lippe River to the north of the coal activity was improved. By the late 1820's the major effort was to seek an even cheaper and more reliable means of transportation, the railroad. First narrow gauge horse drawn railways were introduced. In the 1840's the steam railroad and links from mines to the major lines became the focus of improving transportation.

B. The State and River and Road Improvement

1. The Ruhr River

For the second half of the eighteenth century, the improvement and the free navigation of the Ruhr River appears to have been among the most significant economic and political accomplishments of the Prussian state in the Ruhr region.

Even under Frederick the Great the attitude of the Prussian state toward its splintered western territories was one of discomfort. The problem was what to do with these territories, since the center of the Prussian lands was still in the East. State officials were still somewhat estranged from the different political, economic, and social conditions in these territories. Only gradually did they begin to realize their importance and the significance of the improvement in transportation as a linking element between these splintered territories. It was finally realized that the closer the two areas of Mark and Cleve were connected, the better it would be for their industrial life. They could complement each other. In order to do this, however, certain major technical and, perhaps more important, political obstacles had to be overcome.

In the eighteenth century the future direction of transportation development was symbolized by the non-Prussian city of Mühlheim, located directly on the Ruhr River. The river was navigable from the Rhine to the city. Here the coal was loaded onto Rhine barges and shipped through the small Prussian port of Ruhrort up the Rhine to Cologne, or it was unloaded at Düsseldorf on the Rhine for use by the industry in the non-Prussian Duchy of Berg. Another part of the coal was sent down the Rhine to Cleve and also to Holland.

The city of Mühlheim had become a coal depot. In the seventeenth century several coal dealers had already erected their own depots. The coal traffic was so important to Mühlheim that in the 1740's the dealers of the city formed an association (Ring). It controlled the coal prices and thereby created a coal monopoly which adversely affected some of the regional consumers of coal. This was the case for the chalk industry in the Duchy of Berg. Even harder hit were the industries in the Prussian Duchy of Cleve, since here the wood prices had sharply increased.[3]

Certainly for the Prussian territories, the Mühlheim monopoly created an economic problem. In the Prussian Mark there was frequently an oversupply of coal while in the Prussian Cleve the coal prices were rising, this despite the fact that Cleve and Mark were only separated by a few miles of foreign land. Obviously, the Prussian government was concerned over this state of affairs and became interested in connecting these two separated territories by land or water routes. Adequate land transportation routes did not exist and the Ruhr River down to Mühlheim was not navigable, yet most of the coal mines were located near the river.

As early as October 19, 1734, a Prussian official, Konrad Henning, had been sent to investigate the possibility of making the Prussian part of the Ruhr navigable. He reported it would be technically difficult and costly. In 1737, when the Bochum mining office was established, it was responsible for raising coal production by providing a larger market for the coal. The leading mining official, Decker, had realized the importance of the Ruhr River for the western territories and had expressed it to the General Directory at Berlin. However, Berlin had argued that it would be far too costly to make the Prussian part of the Ruhr navigable. Again in the early 1750's experts from Holland had investigated the Ruhr River and had declared the project feasible.[4]

However, at this time, nearly all the other non-Prussian states owning territories in the Ruhr region were against the improvement of the river. The abbeys of Essen and Werden would not come to terms with Prussia. The Jülich-Berg government in Düsseldorf was openly in opposition; it had received a letter from the Prince of the Pfalz, listing the disastrous effect on the regional economy the opening of the entire Ruhr would have.[5] This letter may have been partially instigated by the prospering city of Mühlheim, which was afraid of losing its central role in the Mark coal market. Basically, however, economics was not the main issue; rather all feared the gradually growing political power of Prussia. In any case, nothing further happened until the 1770's. The same problem also was present in the attempts to make the neighboring river, the Lippe, navigable.

Since by the 1770's the improvement of land transport of coal had also failed, renewed interest was focussed on the water routes. In 1770, a commission sent by Minister Hagen again investigated the Ruhr River. This time the Berlin General Directory agreed with the commission's recommendation to make the Ruhr navigable. However, it asked the local Cleve government to find interested private capitalists, since the state lacked the funds. A group did come together. Among them was a Kettweg textile producer, H. W. Engels, who persuaded the Abbot of Werden to allow navigation on his part of the river. This group started a company of which even the abbey was a member, but by 1774 it had run into financial problems and was dissolved.

The major obstacles were still political. The political objection of the various petty rulers had to be first overcome before the technical aspects of improving the Ruhr navigation could be undertaken. By the 1770's the Prussian government was in the position to exert political pressure. This was especially the case with the Prince of the Pfalz, Karl Theodore, who also was the ruler of the Duchy of Julich-Berg, which controlled a significant part of the Ruhr River. Theodore had to bow to the Prussian pressure. He was just then in need of Prussian support against Austria which was denying his inheritance in Bavaria. Consequently, he agreed to the free navigation of the Ruhr.

Next to the political obstacles were the technical problems. Locks had to be built; the river had to be dammed and forced into a single bed in certain places. In 1774 the first four locks were begun in the Mark. The building of the locks in the non-Prussian parts of the river was very slow. In the cases of the abbeys of Essen and Werden, the building was begun only after a warning from Berlin. The Duchy of Berg waited until the locks were begun in the Prussian part of the Ruhr. Prince George of Hessen-Darmstadt, ruler of the small territory of Broich, claimed he was too poor to build the two locks in his territory. The Prussian government, through the local Cleve government, built it for him and not until 1795 did he repay the cost to Cleve. By 1790, seventeen locks had been built.[6] The Ruhr

River had become navigable for 72 km.

After 1790/92 and until the introduction of the railroad, the Ruhr River was the major transportation route for the Ruhr mining region. Shipment of coal on the Ruhr constantly increased until the early 1840's. In 1816, from a total of 176 mines, 91 mines (35 in the Mark region, 49 in the Essen-Werden region, and 7 in the vicinity of Mühlheim) transported their coal on the Ruhr River.[7] By 1830 nearly 53 per cent of the total Ruhr production was shipped on the river to the ports of Duisburg and Ruhrort. In 1840 coal shipment reached its peak of 58 per cent but by 1850 had declined to 40 per cent of the total coal production. The percentage of course was higher for the Essen-Werden-Mühlheim region and was lowest for the southeastern part of the Mark, where coal was transported by land to the relatively near Wupper Valley.[8]

Not only the total coal shipment but also the traffic increased constantly. In 1820, 3490 ships passed through the locks at Mühlheim; in 1830, 5133 ships; and in 1840, 8633. In 1823 there were 225 ships belonging to 83 owners, most of whom were both coal dealers and mine owners. In 1831 there were 230 ships and in 1838, 377 ships. This was the highest number of ships. From this point on, the ships increased in size rather than number. The average size of the Ruhr barges was 20 tons by the 1820's. By the later 1830's the capacity had reached 75 tons. The ultimate limit imposed by the navigational facilities was reached in the 1850's with 150 ton capacity.

A description of the Ruhr traffic in 1838 will show that shipping on the river had itself become a sizable industry. In 1838, as mentioned above, there were 377 ships engaged in Ruhr shipping. In addition to the 83 owners of these ships there were also involved in the river transport: 1,500 sailors and helpers, 500 horses pulling the barges upstream and 250 horse drivers, 300 loaders and unloaders of coal, and six pilots. This totals more than 2,000 men and 500 horses, at a time when the total mine employment was only about 8,100 miners. Obviously, this in itself made Ruhr transportation costly and would be reflected in the price of coal to the more distant consumer.

Ruhr transportation was by no means dependable and safe. The currents of the river itself were treacherous and changed constantly. Seasonal changes and the weather in general made river transport uncertain. Between 1822 and 1838 the distance between Witten and Mühlheim could at the most be navigated for seven months per year. In the summer the river was frequently too shallow and in the winter it frequently froze. Even under the best conditions the river was difficult and dangerous to use. Ten to twenty ships or 5 per cent of all ships suffered irreparable wrecks during a year.[9]

All this tended to make Ruhr transportation expensive. But another obstacle to the effective lowering of transportation cost on the river were the tolls and levies collected by the state. The Ruhr River, its locks, navigation facilities, and shipment were under state control and the state agency, located at the port of Ruhrort, also controlled the traffic on the river.[10]

A unified system of tolls and levies had been established in 1803, when the entire Ruhr River came under Prussian control. Both the lock tolls and the general levies for shipping on the river were collected in coal. The basic guideline was the ministerial order of March 27, 1803, which came into effect in 1805. It was based on distance to the Rhine: the closer the point of shipment to the Rhine, the higher the tolls. For the same quantity of coal, the toll was, on the average, 13 times greater from the nearest than from the farthest lock on the river.[11] This system re-inforced the _Direktionsprinzip_'s intention that the natural advantages or disadvantages of location in regards to the market should be canceled out.

In 1823, this system was revised and the dues, instead of being collected in actual coal, were now collected in money. Further, a uniform levy was assessed on the basis of barge size. This was the same for all shipments regardless of the distance traveled on the river. However the lock tolls were maintained, although somewhat lowered.[12] By 1839 it still cost a barge of the same capacity 60 per cent more from the lock closest to the Rhine than from the lock farthest up the river.

In 1839, under the pressure of the mining community, the tolls and levies were once more revised and reduced. The lock tolls were completely abolished, and the other levies were to be gradually reduced every two years. In 1849 they were reduced to 48 silver pennies per 16 Zentner of coal. Further reductions in 1852 and finally in 1863 brought them to 20 silver pennies.[13]

After 1803 the whole toll and levy system was continuously opposed by mine owners, coal dealers, and mining offices. They all argued that it artificially increased the price of coal in the Rhine Valley. This burden was particularly resented by the mines in the Essen-Werden-Mühlheim area, which were close to the Rhine.

The toll and levy system was a constant source of conflict between the mining community, including the mining offices, and the state river authority. This agency was formally under the jurisdiction of the Westphalian government and indirectly under the department at Berlin which was responsible for transportation in the state. The Ruhr River authority and the state in general no longer recognized the special interest of the mining community. Both argued that transportation, including water transportation, should benefit the public in general and not just one interest group. The Ruhr authority was interested in all shipment on the river; the Oberbergamt was obviously interested in the coal shipment and therefore found the toll and levies system detrimental to the mining community.

Although the Oberbergamt and the two mining offices had no direct authority over the Ruhr, they were always interested in the improvement of the river and its facilities. They urged the improvement of navigation, especially during the summer months when the water shortage made shipping difficult or impossible. Despite the increasing demand in the Rhine Valley, coal frequently could not be shipped and had to be sold locally at a considerable loss. When obstacles or emergencies arose in river transportation, the mining offices were willing to provide help. This was at least true in the earlier part of the nineteenth century. For instance in 1817 and 1819 certain locks desperately needed repair. The Oberbergamt provided miners for the repair work.[14]

Although Ruhr navigation was continuously improved and canalization was repeatedly suggested by the mining community, no major improvement was undertaken until the mid-1850's, when it definitely was too late. In the 1840's with effective competition from the railways, the Ruhr River lost its dominant role in the expansion of the coal industry.

2. The Rhine River

Before the railroads, the Rhine was the major means of Ruhr coal export to the Rhine Valley cities and to Holland.

During the exceptional years between 1830 and 1835, discussed in more detail in the next chapter, more than one-fourth of total Ruhr coal production was shipped on the Rhine. Nearly all of this coal went to Holland.

In the eighteenth century the Prussian state was interested in the improvement of navigation on the Rhine for much the same reason as for the Ruhr: to unify its territories economically. Therefore the Rhine was constantly worked on. The Prussian part of the river was dammed and canalized where it created difficulties by the constant shifting of its bed. The river was forced into a single bed near Holland in 1763, canalized in 1784 near Wesel and in 1789 near Xanten.[15]

By the end of the eighteenth century the ports of Duisburg and Ruhrort, both at the juncture of the Ruhr and Rhine, had become the major transfer point for Mark coal export. By the beginning of the nineteenth century Ruhrort port was on its way to become the busiest inland harbor in Western Europe. From the very beginning the Prussian state was directly involved in the development of this port. In 1748 the first coal depot was established. Harbor facilities were constantly improved. In 1776 the management of the harbor and its storage facilities came under state authority. The Ruhr coal fund was moved to Ruhrort. The fund was based on tolls and levies from coal transport on roads and river in the Ruhr region, and was to be used for upkeep and improvement of transportation.[16] In the nineteenth century Ruhrort port grew in importance with the improved navigation and reduction and eventual abolition of the many tariffs and tolls on the Rhine. With the advent of the railroad, the port continued to grow in significance for Ruhr coal shipment. It became the major transfer point between railways as well as rail and water.

Although during the first half of the nineteenth century the Rhine was constantly improved, political obstacles and the many tolls and tariffs were a real impediment for bulk good shipments, such as coal and iron. At the Congress of Vienna the Rhine had been declared an international river, with free access granted to all bordering states, and a commission of the bordering states was set up to regulate navigation, but with no immediate results. The Prussian Tariff of 1818 abolished tariffs within the Prussian state and, therefore, on Prussia's part of the Rhine. Nevertheless, in 1830 there still were 34 toll stations between Germersheim, just north of Karlsruhe, and Holland, a distance of 600 km.[17]

There were also tariffs, particularly the tariff by Holland was crucial to Ruhr coal shipment to this country. This tariff and its effect on the Ruhr coal industry will be discussed in the next chapter. Furthermore as long as Holland collected this tariff, the states bordering the Rhine did not have free access to the sea.

In 1831 the Rhine River Commission for the first time came to some agreements regulating the river and some tolls were reduced. Further, in 1834, the Zollverein gradually reduced the tariffs and tolls of its member nations. Not until 1869, however, by the Mannheim treaty, did all the bordering states agree on the free navigation of the Rhine. All levies were then abolished.

The Rhine became the real transport artery after the introduction of the steamship and the iron barges. The first steamship on the Rhine was an English experiment in 1816. Serious attempts came only in the 1830's. During this decade the first attempts were made to form shipping companies with regular passenger and freight service to English, North Sea and Baltic Sea ports. In October, 1837, in his ship "Rhein", Friederich Harkort, the great entrepreneur of the Ruhr mining region, made the first trip to London and his second from Cologne to the Baltic and Oder River port of Stettin. But his company failed. In 1844, a new attempt with governmental backing was made by the Cologne Steamship Company, a joint stock enterprise. Two ships were built and provided successfully regular service from the Rhine to the Baltic ports.

The steamship made large bulk good shipments such as coal relatively fast and inexpensive. With sails, it took a shipment of coal fourteen to thirty days to get from Amsterdam to Cologne; with the new steamship it took only three to five days. According to the Düsseldorf Rhine bridge count of the total number of steamships passing yearly under the bridge, 339 ships had already passed during 1843. By 1845, more than 1,000 ships, and by 1850, nearly 4,000 ships had passed under the bridge.[18]

In the early 1840's the first iron barge with a 150 ton capacity was introduced and by the end of the decade the capacity had increased to 400 tons. In 1850 the first barges with a 500 ton capacity appeared and by the 1860's the normal capacity was 600 tons.[19]

3. The Lippe and Emsche Rivers

The Lippe and the Emscher, both to the north of the Ruhr River, were the other two rivers in the Ruhr mining region. The Emscher River ran in a marshy valley 4 kilometers wide and 25 kilometers long. It was a small river and could not be made navigable without complete canalization. This was eventually accomplished at the end of the nineteenth century.

The Lippe was the only other river within the Ruhr mining region of consequence to coal shipment. It was made navigable between 1819 and 1830.[20] Von Vincke had taken the initiative in the improvement of the Lippe. Already in 1814 he had begun planning and pointing out its benefits for the Westphalian coal industry. The coal from the Dortmund-Hörde area would gain an additional transportation route. On June 6, 1819, the Prussian government granted approval to von Vincke's plan and at the same time provided money from the Ruhrschiffartskasse, the Ruhr River Fund collected from the tolls and levies. By 1830 the Lippe was navigable upwards to Neuhaus, in the far northeastern part of Westphalia.

In 1829 coal transport upstream from Lünen, just north of Dortmund, was 500 tons. By 1832 it had reached 2,900 tons, attaining its peak of 37,000 tons in 1845. Most of the shipment on the Lippe was not coal, however, but manufactured goods from the eastern part of the region. After 1845, coal shipments became negligible because of the Cologne-Minden railway, which passed through Dortmund and ran for part of its route to Minden nearly parallel to the river.[21]

The Lippe River is an especially good example of the lack of a coordinated policy toward transportation by the Oberbergamt. Different interests were represented by the different mining offices. The Mark mining office at Bochum was interested in finding additional transportation for its mines, many of which were not located near the Ruhr River. Particularly by the 1830's mining had moved northwards into the Dortmund area and away from the Ruhr River. Before the advent of the railroad, the Lippe presented a possible additional transportation means. Consequently, the Mark mining office also favored a horse drawn coal railway to the river in the 1830's.

The Essen mining office, however, opposed the use of the Ruhr River fund for the improvement of the Lippe, since its mines did not benefit from the Lippe. It argued that the money should rather be used for further improvement of the Ruhr River. The Oberbergamt located at Dortmund favored the Mark mining office's position on the Lippe. Therefore, even in 1849, it supported the yearly payment of 1,000 thalers from the Ruhr fund for the maintenance of Lippe navigation. The Oberbergamt continued to argue for monetary support even after the Cologne-Minden railway had become operational. It conceded that only a small amount of coal was shipped on the river, but it argued that other mining products benefitted from the Lippe transport. The 1,000 thalers continued to be paid until 1866.[22]

The state's toll and levy policy on the Lippe was essentially similar to that of the Ruhr. Since the cost of upkeep of navigation was higher, the dues on the river were also higher than those of the Ruhr. As in the case of the Ruhr, continual attempts were made to reduce the tolls and levies. But in each case the ministry in Berlin rejected any sharp reduction. Eventually this meant that the Cologne-Minden railway could easily compete with the shipment of coal on the Lippe. For this reason coal shipment lost all significance by the 1850's. Finally in the mid-1860's shipping completely stopped because the mouth of the river at Wesel had gradually silted.

4. Roads

The improvement of land transportation in the Ruhr region represented a long struggle. In the eighteenth century the biggest obstacle was not necessarily political but rather the mentality of the local population which opposed any building or improvement of roads. Opposition was not due to the fear of the loss or division of property since the state did provide compensation. The local population's basic objection was that good roads would not benefit them. In wartime the roads would bring troops for whom they would be forced to provide quarters. They also argued that their income would decrease since bad roads demanded local services. Fresh horses and craftsmen for repair were constantly needed. Travelers had to use the lodging facilities due to the road and the weather.

Therefore, the local population did not willingly contribute time and effort in the maintenance of the roads. Further, because of this opposition, even when new roads were to be built, they were frequently routed through uninhabited and desolate country and not through the populated valleys.

Despite these transportation difficulties, coal export by the 1750's had increased so much that a coal scarcity appeared in Mark. This was mostly due to the industrial development of the southern hill region of the County of Mark. In a report of September 30, 1752, the areas of coal export included the eastern part of the non-Prussian Duchy of Berg, especially the area of the Wupper Valley, south of the Mark coal region. Here industrial enterprises flourished, particularly textile in the cities of Barmen and Elberfeld. Here the Ruhr coal was in demand because of its good quality. Within Prussian possessions an export duty was imposed on the Mark coal to ensure a supply for the southern hill region of the Mark itself. But soon this demand was met and again export was desirable. The development of a transportation system became crucial. In general, by the 1760's there had been little progress in road building since medieval times.[23]

In the 1760's a land route was proposed. The idea came from J. G. Müser, a mine owner and later a mining official. The Gahlen Coal Road was to be built by the local population. For this purpose 19,000 thalers were borrowed from the Mark Mining Fund, to be repaid with the profits from the coal sales.

The Gahlen Coal Road was intended to break the Mühlheim monopoly. The road was to circumvent Mühlheim and connect Mark and Cleve. The route went from the Ruhr River by Welper, through Hamme near Bochum to the Emscher Valley, to Dorsten and ended in Gahlen on the river Lippe, which was navigable at this point down to the Rhine. But the building went slowly and the project soon lacked the necessary finances and the construction stopped. The state was unwilling to provide more funds. This was at a time when the Cleve government was desparate for coal because of severely rising prices.[24] It was even forced to allow non-Mark coal to enter.

Finally, between 1788 and 1792, under the urging of Stein, two major highways were built.[25] These two highways connected the region internally and linked it with other regions. They provided for the first time a fairly easy means to transport coal to some of the centers of consumption in or near the Ruhr region. One of these linked the Rhine territories with Berlin. It began in Düsseldorf and ran through the Wupper Valley to Hagen, cutting from there through the eastern part of the coal region to Soest eventually reaching Berlin. This road led through the most populated area of the Westphalian industrial region and a major coal market. The stretch from Hagen to Wupper Valley cities of Barmen and Elberfeld, known as the Enneper Road, was paved and in fairly good condition. This Enneper Road was the major coal transportation route for the mining areas south of the Ruhr River.

The second highway went from Frankfurt to Holland. It passed through the Siegerland and came up the Lenne Valley. For a short distance it joined the other highway until Hagen. From Hagen, it passed through the Ruhr mining region, from Witten on the Ruhr River through Bochum, Essen, and northwards to the Lippe city of Wesel, where it followed the Rhine to Holland.

Although additional roads were built, the two highways remained the most important roads for coal transport even by the 1820's. Gradually during the rest of the first half of nineteenth century a network of secondary roads were built using the two highways and the river Ruhr, Lippe and Rhine as the basic structure.

On all of the roads, state or private, tolls had to be paid. The tolls on the coal roads, as most of the secondary roads were commonly called, were particularly high. The result was that throughout the first half of the nineteenth century tolls and inadequacy of land transport still contributed greatly to large local price variation.

Since the Oberbergamt felt responsible for the sale of coal, it was also interested in the improvement of land transport. Again the mining administration lacked direct authority and again it had no unified, coordinated program influencing the improvement of land transport. It encouraged and even provided technical help in the building of some roads, but it discouraged others.

After 1815 some of the initiative for the construction of roads came from the provincial government and its president von Vincke. In the decade following the Napoleonic Wars, von Vincke encouraged private enterprise to contribute to the building of a badly needed network of roads. The state would not burden its depleted treasury with road building.

For the first time joint-stock companies made their appearance in the Ruhr mining region. Von Vincke tried to encourage these private and semi-public companies in the building of roads by personally participating in each project. He even used his private wealth to buy shares.[26]

Normally the coal roads were built by joint-stock companies in which mine owners, coal dealers, and interested communities participated. The Mühlheim-Borbeck road and the Steele-Rellinghausen-Werden road, located in the middle of the coal producing areas, were built by joint-stock companies. The mines and the towns bordering these roads participated in their construction and later in their upkeep. The Mühlheim-Borbeck joint-stock company had a capital of 30,000 thalers and was heavily represented by the Mühlheim coal dealers.[27] In these cases, the _Oberbergamt_ gladly provided technical assistance in the planning and even provided some loans from the mining fund. It felt that these roads were absolutely necessary for the continued sale of coal from the group of mines involved.

C. The State and Railway Development

The greatest improvement in land transport in the nineteenth century came with the introduction of railways. The major objective of private enterprise was to introduce new modes of land transport, first the coal railways and later the steam driven railways themselves. Both of these means were not only more rapid and dependable, but more versatile and cheaper than road and water transport. The actual building of railroads came rather late in Prussia compared with England or Belgium. However, the relatively late building in the 1840's had been preceded by an active planning stage. During the late 1820's and all of the 1830's much of the initial opposition and conflict of interests were resolved. Nevertheless, until 1838, the Rhenish and Westphalian provinces were left without a railroad. The continuous conflict of interest both among private interest groups and the state kept the first major railway lines from being completed until the late 1840's.

1. General Prussian Policy

The historical development of this important transportation industry[28] can be divided into three periods. Before 1847 it was private initiative which instigated railroad building and private enterprise owned the railways. Between 1847 and 1873, the state gradually entered and the period is usually known as that of the "mixed railway system." The state and private enterprise shared in the building and ownership of the railways. After 1873 a gradual nationalization of all Prussian and other German railroads took place.

In the early 1830's, when private interests were actively promoting the railway idea, the king, Frederick Wilhelm III, reflected the opinion of many in showing a total indifference. The king's often quoted statement at the suggestion of a Berlin-Potsdam rail line serves as an example. "I cannot imagine gaining any great pleasure from arriving a few hours earlier or later in Potsdam." But at the same time his son, the crown prince and future Frederick Wilhelm IV, was a railroad enthusiast. And his equally often quoted statement was to symbolize the future state involvement. "These railcars, rolling through the world, no man will be able to stop."[29] Subsequently, when he became king in 1840, he actively pushed for a planned railway system with Berlin as its center.

Prussian state policy toward railroads was reflected in the series of laws and orders of the king which established the pattern of state intervention. The first law of November 3, 1838, was an attempt by the state to provide a future legal basis for influencing the building and operation of railways. The law was vague and could be interpreted on many levels, but it gave the state wide powers over railway companies. Every railway joint stock company had to obtain a concession.[30] A cabinet order was issued on November 22, 1842, declaring that the state would in the future be prepared to assist in the building of railways by guaranteeing the interest on shares at 3 1/2 per cent on approved projects. The state also could buy one seventh of the stock of a company. This had already been mentioned in the 1838 law, but now it was restated. Another cabinet order of April 28, 1843, established a railway fund to be used by the state to assist railway building. It also provided means for the state to own the controlling shares of whole companies. The government purchased some shares; other shares and loans received a state guarantee of interest. The money spent on guaranteeing interest on the shares eventually was recovered.[31] The prosperous railways had to contribute to the fund and this money was then passed on to those investors having shares in less successful companies.

The order of May 24, 1844 to the Berlin stock exchange in essence prohibited certain important stock exchange operations, particularly affecting railway shares. According to this order all trading of railway shares had to be registered and approved by the finance minister. The law of December 7, 1849, passed by the Diet which had been reconvened in 1847 after a lapse of 27 years, granted the state the right to build railroads. Now the way for a state-owned railway system was cleared.

But before 1847 the state did very little to encourage or even directly help in the building of railroads. The state by 1847 had contributed only five to six million thalers as compared with one hundred and fifty million thalers from the 26 Prussian railway companies.[32] The state's guarantee of 3 1/2 per cent on shares did not attract capital because the general existing rate on all other businesses was 6 per cent.[33] In addition the state nearly wrecked all positive efforts on behalf of railroad building with the order of May, 1844, requiring the registration and approval by the finance minister of all trading in railway stock. Furthermore, there were strict controls over the railway joint-stock companies. Concession and land were necessary for joint-stock companies, and only the king himself, not the bureaucracy could grant them.[34]

Nevertheless, some of the leading civil servants were active promoters of early state participation in railroad building and of eventual state ownership. One of the strongest advocates for state railways was the Minister of Commerce, August von der Heydt (1848-1862). Under his urging the mixed railway system of the same period came into existence. Von Heydt felt that the young industrial sector was too small to provide the large amounts of fixed capital and argued that private ownership would be unprofitable.

In actuality, from the very beginning, there was a fusing of interest in the building and ownership of the railroads in Prussia. For the boards of the railroad joint-stock companies "the bourgeoisie selected prominent aristocrats, although often only as honorary members.[35] This was done to gain the concessions for the companies and approval of the specific project so building of a railroad could get started. This frequently took several years. The aristocratic shareholders, often princes and barons, frequently resold their shares to business houses.

In general, the state and private interests had joined

in the building of railways. The railroads very early took on the nature of a public utility. They depended heavily on the state for engineers and trained managers simply because these did not exist except in the employment of the state. These younger sons of the aristocracy had received an education.

The state's political interest in railroad building was the hope that the railroad would provide the means of unifying the Zollverein. To paraphrase Friederich Luetge, the Zollverein attempted formal economic integration in 1834, but the railroads provided the real means to the economic integration of Germany, to be crowned by the actual political integration in 1870.[36]

2. The Direktionsprinzip: Coal Railways

The attitude of the Prussian state toward transportation in the Ruhr mining region was reflected in its exercise of its right of eminent domain, the condemning of private property for public use. The state used this as a major policy instrument, influencing transportation development, especially the coal railways.

Before the introduction of railroads, the expropriation right of the state had been important only in the building of roads. Obviously roads were essential for the transportation of coal where immediate access to water did not exist. Further, even where major public roads existed, mines still needed access to them. The 1766 codification and the mining section of the Allgemeine Preussische Landrecht of 1792 stated that the property owner was only required to provide that portion of land which was directly necessary to the actual operation of a mine. The mining law did not explicitly mention land necessary for access to the mining property and for storage near major transportation routes. This land for access and storage became the source of constant legal conflicts between mine and property owners.[37]

Finally, the Oberbergamt pressed the responsible department at Berlin to clarify this situation. The result was the declaration of October 27, 1804. Henceforth the property owner was obliged to provide the necessary land for the construction of access roads and storage facilities, but two qualifications of later significance for railroads were added. The first required a proof that the land requested was absolutely necessary for the continuing operation of the mine. It was not sufficient to simply indicate a need; the operation of the mine had to depend on the construction of such a road or storage facility. The second qualification was vaguely formulated. It stipulated that "another business activity, just as beneficial to the welfare of the state and public, would not suffer" from the building of such a road.[38] In 1804, this declaration truly favored the mining interests, but in the 1830's it was the basis for the government's opposition to the building of horse drawn railroads and in the 1840's to the construction of links by mines to major rail lines.

Until the 1830's the Oberbergamt together with the provincial government determined the necessity of the land in question and also arbitrated any conflict. The Oberbergamt handed in a detailed report and its recommendation in each case. The provincial government normally gave its approval, unless property owners indicated extreme opposition. Formally in each case, however, the responsible department at Berlin had the final approval. But normally no basic conflicts arose before the 1830's. If there was disagreement between mine owners and property owners and a legal decision became necessary, the law required the payment of all legal fees by the mine owner.

The real conflict between the mining interests and the state arose with the private construction of horse drawn coal railways and the coal links to major railroads. In both cases construction required not only land but frequently a government concession for the formation of a joint-stock company. The 1804 directive remained essentially in effect until the 1850's. The 1838 railway law did not clarify the right of mines to build private railway links. In 1844, the Oberbergamt still tried to get clarification from the finance ministry. It received as an answer that the 1838 railway law did not give mines the right to build railway links, "as long as these links were only advantageous to a particular mine or mines and did not benefit the general public transport."[39] Nevertheless by the late 1820's narrow gauge horse drawn coal railways were being constructed. In 1825, Friederich Harkort was already advertising the future importance of railways for the western Germanies. Some mining officials listened to him. In 1828 Harkort took the lead in the establishing of the first German joint-stock railway company. His company built a narrow gauge coal railway in 1830.

Heintzmann, a leading mining official of the Essen _Bergamt_, inspected Harkort's experimental "Palmersche Bahn." He was impressed and managed to convince the _Oberbergamt_ to allow him to plan a railway from Steele on the Ruhr River to the Wupper Valley city of Elberfeld. Immediately a conflict of interest arose. The Bochum mining office opposed the plan. It argued that the mines in its administrative district, further up the Ruhr River, would lose some of their market, although it agreed in principle that this railway would lower transport cost. At the same time the Essen _Bergamt_ favored this project since the mines in its district would benefit.[40]

This disagreement and rivalry among the mining offices resulted in the _Oberbergamt's_ withdrawal from the project. In the future it left the early building of coal railways to private initiative. Nevertheless, by 1837 a total of thirty kilometers of coal railways were in existence. All had been built by private enterprise and the typical form had been the joint-stock company. Twenty coal railways with a total length of 13 kilometers were in the Essen-Werden region. Seventeen railways with a total track length of 17 kilometers were in the Mark region.[41]

These coal railways usually operated in side valleys of the Ruhr River. They either had the Ruhr River or the Enneper highway as their destination. The Muttenthal railway served a group of mines between Bommern and Herbede and even crossed the Ruhr River at Witten. The Deilbahn went from Nirenhof south of the Ruhr River to the river. There were a few other major coal railways: the Sellerbeck railway from the Sellerbeck mines to Mühlheim on the Ruhr, the Neuermisgunst railway south of Bochum to the Ruhr, and the Brockhausen railway from Weitmar to the Ruhr. Both the Deilbahn, which in the late 1830's was extended and renamed the Prince-Wilhelm railway, and the Sellerbeck railway, became eventually part of the later major steam driven railways.[42]

For those mines which already had access roads it was difficult to prove the necessity for the construction of coal railways or railway links. As long as the mine benefitted from the existing roads and was able to continue operation, it normally did not have sufficient grounds to demand additional land for a railway route. This remained the case despite the fact that it was generally recognized that non-railway transport was more costly. For example, in the early 1830's, the Schlebusch-Harkort coal railway could not take a direct route to the Enneper highway but had to take a more difficult and costly route. The Schlebusch ore mines had not been able to provide sufficient grounds for expropriation procedure. They already possessed a coal road built in 1808.[43]

During the 1840's some of the leading mining officials of the Oberbergamt and the Westphalian Bergamt at Bochum favored the building of rail links for some of the mines in the Mark region. The Finance Ministry denied the requests; it was not willing to give special privileges to the mining sector. It argued that it saw little reason why private railway connections should be favored over other transportational means. Consequently, the Oberbergamt was reprimanded in 1844. In the future it was to "act according to the established law and not give its own interpretation, favoring the mining interests."[44] The established law rested basically on the 1804 directive applying to the building of access roads.

The mining administration did not necessarily favor the general building of coal railways or railway links, except in those cases where it felt whole groups of mines would benefit because other means of transportation were lacking or extremely costly. Both the Ministry and the Oberbergamt agreed, however, that they did not want every mine to have its own connection. When Stinnes wanted to form a joint-stock company to build a coal railway from his shaft mine Graf Beust to the Ruhr River city of Mühlheim, the Ministry definitely discouraged him. It only gave its permission when Stinnes declared he would build the railway on his own account and promised not to form a joint-stock company.[45]

The mining administration basically feared that indiscriminate building of coal railways and links would result in giving some mines advantages over others, or result in disastrous competition. In addition, sometimes conflict of interest between the mining offices resulted in hindering the building of private railways.

3. Railroad Development

The history of the construction of the first major railways in the Rhenish and Westphalian provinces reflects the energetic initiative of the private entrepreneurs, conflict of interests, regional rivalry and jealousy. It also reflects the attitude of the Prussian government toward the railroad.[46]

In the late 1820's in the Rhineland and Westphalia, several industrialists and merchants promoted the construction of railways. Friederich Harkort, the brothers August and David von der Heydt, Rudolf Camphausen, and David Hanseman were the leading promoters. The major promoter for the idea of a Rhine-Weser railway was Friederich Harkort.

This idea received general support from all interested parties. Harkort also found strong support for his scheme in Finance Minister Motz. Motz saw in this railway a means to make the Rhenish-Westphalian industries less dependent on the Rhine, whose mouth was controlled by Holland, which had, in 1816, instituted a damaging tariff against Prussian goods. Motz was so enthusiastic that he immediately ordered the presidents of Westphalia and Rhineland to provide the technical personnel for the surveying of the route. The Oberbergamt also favored the project, hoping to gain an additional means of marketing the increasing coal production.[47]

Motz unfortunately died in 1830 and with him the project came to a complete halt. In addition, Holland by the end of the year had considerably reduced its tariff on the Rhine for Prussian goods. Furthermore, Interior Minister Schuckmann, under whose jurisdiction transportation fell, was not a railway enthusiast. To him, "the technical difficulties were too great."[48] He rather favored the building of canals. In 1833, in a letter to von Vincke, he even prohibited the use of state personnel helping the technical planning of the proposed railway.[49]

Meanwhile the Rhine-Weser project had come before the Westphalian Diet in 1831, and was accepted by a majority vote. The Diet asked the king for permission to build the Rhine-Weser Railway from state funds. If this were not possible, the Diet advised the state to provide one-half of the shares without demanding dividends for ten years. On July 31, 1832, the request was turned down by the government, but the state held out some hopes. It hinted that if a company built the railway the state would be prepared to buy some shares. But in the following year the Diet requested that the state guarantee at least 4 per cent interest on shares. This too was rejected.

When a joint-stock company did begin the building of the Rhine-Weser line from Minden to Unna, which eventually was to have passed through Dortmund, Witten, and Elberfeld to Cologne, the Finance Ministry even prohibited the *Oberbergamt*'s participation in four or five shares in the company. Construction, which had begun without any state help, came to a halt by 1838. Particularly the Mark *Bergamt* lamented this failure because the proposed line would have passed through the coal rich area of Dortmund-Hörde. However, in the same year a joint-stock company, later a part of the Berg-Mark Railway, had, without state assistance, completed the Dusseldorf-Elberfeld Railway.[50]

In the 1840's the state did change its policy toward railroads in the western provinces. It provided some important financial help. In 1842, the state asked a commission representing all the provincial Diets to give their opinion on the future role of the state in the construction of railroads. The commission answered in the affirmative the state's inquiry, "whether the construction of a substantial railway net was a real need for the country and whether the state should encourage railway building by guaranteeing the interest on private capital in such projects."[51] It should be remembered that the Prussian Diet had been dismissed by the king in 1820, and was not recalled until 1847.

In 1844 the Cologne-Minden Railway Company began construction. The state contributed one seventh, or 1,800,000 thalers, and guaranteed a 3 1/2 per cent return on the private capital. In the case of the Berg-Mark Railway from Elberfeld to Dortmund (begun in 1843), the state contributed one fourth, or 1,000,000 thalers, and waxed its rights to dividends until the private capital had received its 3 1/2 per cent.[52] It is interesting that 10 years earlier the Westphalian Diet had asked for similar aid and was rejected.

In the planning of the first railways, three major sectional interest groups clashed. The principal conflict was over the routes. This was especially the case in the long discussed Rhine-Weser project. The immediate result of this conflict was the postponement of construction. In addition, this opposition resulted in the establishment of four major companies, which were to remain the most important railway companies in the Rhenish-Westphalian industrial area. The four companies were the Cologne-Minden company, the Berg-Mark company, the Rhenish company, and the Prince-Wilhelm company. They received their state concessions in 1843, 1837, 1837, and 1845 respectively.

The leading promoters representing the sectional interests of the Wupper Valley and Dusseldorf were Friederich Harkort and August von der Heydt. Rudolf Camphausen represented the Cologne interests and David Hansemann represented the interest on the left bank of the Rhine.

Harkort and von der Heydt advocated the construction of railways in the industrial districts on the right bank of the Rhine and in the valleys of the Ruhr, Wupper, and Sieg. They argued that manufacturers of the Rhineland and Westphalia would be able to obtain coal from the Ruhr mining region at reduced cost. For instance, in 1834, a report of the Elberfeld Chamber of Commerce drew attention to the high cost of Ruhr coal. The cost made it difficult for the industrialists of the Wupper Valley and other parts of the western provinces to compete with their British rivals. This report stated that English metal buttons were being sold in the Wupper Valley at a lower price than buttons made locally. This was despite the fact that English manufacturers paid higher wages and had no advantages as far as raw materials were concerned. Furthermore an import duty was levied on metal buttons entering Prussia. "The main advantages enjoyed by English button makers," according to this report, "is cheap coal."[53] And according to Harkort and von der Heydt, the major reason for cheaper coal was lower transportation cost.

Rudolf Camphausen, representing the Cologne interests, was a member of the prestigious Cologne City Council and a corn merchant. He was the leading member of a committee set up in Cologne in 1833, promoting the construction of a line to the Belgian frontier. For Camphausen the main purpose of the construction of such a line was to handle through traffic between Antwerp and Cologne. David Hansemann, the head of an Aachen insurance company, argued that the railway should be constructed primarily to promote the industrial prosperity of the territories on the left bank of the Rhine.

Political rather than economic factors appear to have been the cause for the delay by the Berlin authorities in granting a concession for this route. In 1830 the Belgian revolution took place and the Prussian government feared that Belgian liberal views would contaminate the Rhinelanders if the Cologne-Antwerp line were completed. In 1837, the government finally decided that the companies of Camphausen and Hansemann were to be amalgamated. In 1837 the new Rhenish railway company received its concession. By 1843 the first main line was completed after the government had bought some shares in the company. This railway successfully competed with shipping on the Rhine.[54]

In the case of the Rhine-Weser project, the conflict caused by the proposed route was between Hansemann and von der Heydt. Hansemann proposed that the line should run through the Ruhr region from Duisburg following the Emscher Valley to Dortmund, Hamm and Minden. Von der Heydt favored the shorter but costlier route from Dusseldorf through the Wupper Valley to Dortmund and then to Minden. This disagreement resulted in the formation of two separate companies, the Berg-Mark railway, led by Von der Heydt, and the Cologne-Minden railway. Part of the Berg-Mark line from Dusseldorf to Elberfeld had already been built by 1838; the rest, from Elberfeld through Hagen, Witten to Dortmund where it would meet the Cologne-Minden railway, was completed by 1848. This railway linked the industrial valleys of the Wupper and the Ruhr.[55]

At the time of its construction, the Cologne-Minden railway ran north of most of the mining activities in the Ruhr region. But since mining was also moving north during the 1840's, this line was to benefit the shaft mines directly. Construction began in 1843. In 1845, it reached Dusseldorf, Duisburg in February, 1846, Hamm in May, 1847, and finally Minden in October, 1847.[56]

The last major railroad of significance for the Ruhr industrial area was the Prince Wilhelm railway. It received its concession in 1845 and was completed in 1847. The railway included earlier coal railway, the Deilbahn. Its route was from Steele on the Ruhr River to Vohwinkel, where it joined with the Berg-Mark railway just west of Elberfeld.

These four major lines became the basis for the network

which was gradually added after 1848. In 1862 the Berg-Mark railway finally completed a line joining the Ruhr region cities of Dusiburg, Mühlheim, Essen, Bochum, and Dortmund. Now all the major cities in the region had railway connections. After 1848 the state also entered the railway industry directly by taking over the management of the Berg-Mark railway on October 15, 1850, and the Prince Wilhelm railway in April, 1854.

As early as the mid-1840's some mines benefitted from the building of these major rail lines. The mining administration in the Ruhr region, still committed to uphold the <u>Direktionsprinzip</u>'s functions, became concerned about mines which were unfavorably located in regards to these railways. Both the private mine owners and the mining offices were interested in links to these major lines. The mining offices were interested in promoting railway connections for entire groups of mines. But the finance ministry would not participate or allow the mining offices to participate in such projects. Using an 1804 cabinet order, it argued that "such branches were not of sufficient general public importance to warrant the state burdening itself with such risks."[57] In the 1840's the state had been giving encouragement and financial help to the major rail lines; it still did not want to encourage the construction of coal railroads and branches linking major lines. The state did, however, lead the private joint-stock companies to expect the concession and the condemnation of land necessary for the right-of-way.

The only exception which the state made to its policy was in the case of the Prince Wilhelm railway, which was predominantly a coal railroad. The mining offices had been able to convince the Finance Ministry that this railway had more than local significance. The mining administration was allowed to contribute 25,000 thalers from the mining fund.[58]

In general, by the 1840's the mining offices, although prohibited from giving direct help, did encourage the building of railway links. For one thing, the major rail lines were an accomplished fact and would affect the Ruhr region's coal sales regardless of the mining administration's attitude toward railways. Furthermore, the mining officials had also begun to understand that through the railroads "the sales of coal increased and the price for the more distant consumers was actually lowered."[59] The mining offices even attempted to resolve private disputes about railway links. Finally, in 1844, when there was a great deal of unemployment in the mining sector, the mining administration even tried to find temporary jobs for miners in railway construction.

The mining industry did benefit from the construction of the major lines, but the full impact was not immediate. The railroads were after all a new means of transportation and were not immediately ready for bulk goods shipment. In general, the first railroads had a more immediate and direct impact on passenger service. In the early period of railroads the largest share of income was from passengers. Facilities for large freight shipment simply did not yet exist. Facilities for loading and unloading at cities and ports had to be built. It simply was easier to ship people than coal and iron ore. Furthermore, a unified rail network did not exist. Each company had its own policies. There were no uniform freight rates.

Eventually rail freight charges were structured similarly to those of river transportation. Distance was the essential basis but the value of the good was also assessed. Initially, there was a flat charge per mile. The low value, long distance goods, such as coal and iron ore, did not benefit. No special provision was made for long distance shipment of bulk goods.[60]

The intent of rail tariff reform in the late 1840's and particularly throughout the 1850's was to secure not only a uniform tariff system but also special treatment for long distance shipment of bulk goods. Finally in the 1850's some special rates for certain goods were introduced, allowing a reduction in charges the further the goods were sent.[61]

There were, of course, other problems which this young industry had to solve before it had its full impact on the shipment of industrial goods. A uniform Prussian tariff system was not achieved until the 1860's, although some of the major railroads made agreements earlier. Transport did not always take place on the shortest possible routes because of the lack of cooperation among the railways. Many of these problems were solved after 1848, during the "mixed railway system," when the state gradually took control of many privately owned companies.

FOOTNOTES - CHAPTER III

1. Henderson, p. 48.

2. Additional information may be found in the following: V. Kern, Die Wirtschaftliche Entwicklung der Ruhrschiffahrt (Diss. Köln, 1921). (mining Libr. Essen) W. Kliche, "Die Schiffahrt auf der Ruhr und Lippe im 18. Jahrhundert," Z.d.Bergischen Geschichtsvereins, (1904) (mining Libr. Essen); G. Strotkötter, Die Bestrebungen zur Verbesserung der Schiffbarkeit der Lippe im 15., 17. und 18. Jahrhundert, (Münster, 1895) Münster, 1895) Münster Archive; G. Strotkötter, Die Lippeschiffahrt im 19. Jahrhundert (Münster, 1896) Münster Archive; Der Rheinstrom und seine Nebeuflüsse, ed. Centralbüro für Meteorologie und Hydrographie im Grossherzogthum Baden, (Berlin, 1889) (Münster Archiv).

3. Spethmann, Ruhrgebiet, p. 138, 139.

4. Spethmann, p. 140.

5. Spethmann, p. 141, Pounds, The Ruhr (Bloomington, 1952), p. 52.

6. The above two paragraphs have been based on: Spethmann, pp. 144-150.

7. B.A. Bochum, Bericht vom 19. 7. 1816, (OBA-37/12).

8. Reuss, "Mitteilungen...", p. 14; A. Adolphe, Ruhr kohlenbergbau, Transportation und Eisenbahntarifpolitik, p. 93.

9. The above two paragraphs are based on: Spethmann, pp. 156-158.

10. Achenbach, "Geschichte...", p. 240.

11. Regulativ zur Schleusengelderhebung vom 31. 5. 1805, (Staatsarchiv Münster) Oberpräsidium Münster, B 972.

12. Achenbach, p. 240.

13. Krampe, pp. 174, 175, 204.

14. Krampe, p. 170 based on O.B.D. 145, Blätter 47, 88, 91, 95.

15. Spethmann, Ruhr, pp. 151, 152.

16. Spethmann, p. 153.

17. Adolphe, p. 93.

18. Spethmann, p. 310.

19. Adolphe, p. 94. The first iron barge was introduced in 1841.

20. Strotkötter, Lippeschiffahrt, pp. 5-9.

21. Krampe, Staatseinfluss, p. 117.

22. Krampe, pp. 177-178.

23. Spethmann, pp. 137, 138.

24. Spethmann, p. 143.

25. Spethmann, p. 154; Henderson, p. 57.

26. L.V. Vincke, "Bau der Kunst strassen durch Privatvereinigung," Hermann (1816, Nr. 15). (mining Lib. Essen).

27. Krampe, p. 185.

28. Usefull works on the history of railroads: Fritz Krönig, Die Verwaltung der Preussischen Staats-Eisenbahnen, (Breslau 1891); K. Kumpmann, Die Entstehung der Rheinischen Eisenbahn gesellschaft, (Essen, 1910); Carl Heuser, Canäle und Eisenbahnen und ihre wirtschaftliche Bedeutung, (Berlin, 1880); F. J. Pieler, Das Ruhrtal Reise auf der neuen Ruhrtal-Eisenbahn mit Ausflügen in die Umgebung, (Arnsberg, 1871); Hans Baumann, Deutsches Verkehrsbuch, (Berlin, 1931); Emil Richter, Die Entwicklung der Verkehrgrund lagen, (Berlin, 1878); F. Weichsel, Die Rhein-Weser Eisenbahn A.G. (Leipzig 1838) and a particularly well documented history: Arthur von Mayer, Geschichte und Geographie der deutschen Eisenbahnen (Berlin, 1891).

29. Mayer, p. 7.

30. Mayer, p. 106; Dietrich Eichholtz, Junker und Bourgoisie vor 1848 in der Preussischen Eisenbahngeschichte, p. 87.

31. Mayer, pp. 106, 107.

32. Eichholtz, p. 112.

33. Eichholtz, p. 112; Mayer, p. 111.

34. Eichholtz, pp. 115-119; Mayer, pp. 108-111.

35. Eichholtz, p. 116.

36. Friedrich Lütge, <u>Deutsche Sozial-und Wirtschaftsgeschichte</u>, (Berlin, 1952), p. 175.

37. <u>Codification of 1766</u>, Article 72, <u>Landrecht</u>, Part II, Article 16, SS 109.

38. "Declaration vom 27. 10. 1804 wegen über lassung des Grund und Bodens an die Bergbau treibenden Gewerke zur Anlage der Abfuhr wege und Niederlagen," printed in C. Karsten, <u>Die Revidierte Bergordnung 1766</u>, p. 101.

39. Krampe, p. 167.

40. Krampe, pp. 186-187.

41. BA Essen, übersicht der auf den Zechen im Essen-Werden schen BA-Bezirk gelegten eisernen Schienen, OBA-38/4; BA Bochum, Tabelle and Aufstellung von den bis März 1836 in den Märkischen Revieren vorhandenen Eisenbahnen, OBA-38/4; cited in Krampe, p. 187

42. Mayer, p. 111.

43. F. Horkort, <u>Geschichte des Dorfes, der Burg und der Freiheit Wetter</u>, p. 32.

44. Finanzministerium an Oberbergamt 9. 6. 1844 (OBA-38/5) cited in Krampe, p. 67. n.

45. Krampe, p. 168, n.

46. Mayer, pp. 102-115.

47. Krampe, p. 189.

48. Waldeck, "Die Entwicklung der Berg-Märkischen Eisenbahnen," <u>Archiv für Eisenbahn wesen,</u> (1910), p. 591.

49. <u>Ibid</u>., pp. 591-593.

50. Krampe, p. 190.

51. Bessel-Kühlwetter, <u>Das Preussische Eisenbahnrecht</u>, (Köln 1855-1857) Vol. II, pp. 4-5; Krampe, p. 190.

52. J. Michaelis, <u>Deutschlands Eisenbahnen</u>, pp. 104-106, pp. 40, 374; Mayer, p. 359.

53. Henderson, p. 152 n.

54. Mayer, pp. 470-472.

55. Mayer, pp. 230-232.

56. Mayer, pp. 466-68.

57. Krampe, p. 167.

58. Krampe, p. 192.

59. <u>Ibid</u>.

60. Adolphe, p. 112.

61. <u>Ibid</u>.

CHAPTER IV

STATE ADMINISTRATION AND PRIVATE ENTERPRISE: CONFLICT AND GROWTH IN THE FIRST HALF OF THE NINETEENTH CENTURY

Introduction

This chapter analyzes the tension which developed in the course of the first half of the nineteenth century between state policy and private enterprise. The growing industry and changing economic conditions in the Ruhr region, on the one hand, and the unwillingness and inability of the state and its administrative apparatus to adopt a more flexible policy toward new demands, on the other, created tension. More specifically, the conflict centered on the mining administration's attempt to enforce the *Direktionsprinzip* while private entrepreneurs attempted to change or circumvent this complex and restrictive system. Consequently private enterprise and a conservative, bureaucratic-minded state administration were in continuous conflict.

The healthy annual rate of growth of coal production of more than 5 per cent between 1815 and 1850 was essentially a credit to private ingenuity and enterprise. As Von Vincke, the president of Westphalia, argued, the *Direktionsprinzip* may have been appropriate in Stein's day, but no longer suited conditions after the Napoleonic Wars.[1] Many state officials and private entrepreneurs felt that economic conditions in coal mining had changed sufficiently that the time had come to withdraw the tutelage of the state from this industry.

In the nineteenth century the Ruhr mining region no longer received favored or special treatment from the government as it had in the eighteenth century. Furthermore, after 1815 Prussia reorganized the administration of its western territories, dividing them into the two provinces of Rhineland and Westphalia. But what is of significance for the Ruhr mining region, the entire administration of the territories was now no longer in the hands of one official. Stein as the highest political and mining administrator had been able to coordinate mining, transportation, and political administration. Ludwig von Vincke, the highest political official in Westphalia, had no authority over the mining administration and very little over transportation. Consequently, the *Direktionsprinzip* policy no longer served as the guide for a general economic policy for the development of the entire Ruhr region. It was relegated to a limited mining policy.

Finally, while in the eighteenth century the state had taken the leading role in the development of the industry and region, in the nineteenth century private enterprise increasingly took the initiative. In the case of transportation, discussed in the previous chapter, private enterprise had successfully taken the leading role, particularly in railroad construction, and eventually forced the state toward a new attitude and policy during the 1840's. In the mining industry a similar and related process occurred. Private entrepreneurship was taking the initiative and gradually pressured the state to adopt a new attitude and policy toward the industry. During the 1840's but primarily in the early 1850's the state changed its policy. This change in policy and the corresponding legal reforms will be discussed in the next chapter.

In the meantime, chiefly after 1830, tension and frequently open conflict occurred between the mining administration and private enterprise; this was especially visible in the administration's attempt to achieve its basic objectives. The rest of this chapter analyzes this tension in terms of the mining administration objectives: orderly administration of the law, regulation of production and marketing, and control over individual mining operations and business organization.

While the administration was able to maintain certain aspects of the *Direktionsprinzip*, in others it was either forced to relinquish or give a more lenient interpretation and enforcement. In general, during the 1840's the state and its administration adopted a more liberal attitude in the application of the *Direktionsprinzip*.

A. Production Controls and the Shaft Mine

During the first half of the nineteenth century shaft mining, particularly the penetration of the water-bearing layer after 1830, was advanced at the initiative of a small group of profit seeking and farsighted entrepreneurs. The large scale shaft mine came into existence despite the opposition of the mining administration. Between 1815 and 1850 the administration still had as an objective the regulation of production by controlling the number and size of firms in the industry. It successfully controlled the number and attempted to control the size of the firms through the complicated procedure for gaining entrance into the industry. It tried to control mine size by discriminating against the shaft mine. Although the first shaft mine did not penetrate the water-bearing rocks, the administration feared the competitive effect on the structure of the industry and its own loss of control over the industry.

The mining administration in the nineteenth century still appeared to be guided by the assumption that the most suitable operating units compatible with the *Direktionsprinzip* policy objectives were relatively small firms of similar structure and employing similar techniques. Consequently, the new shaft mine presented a problem. In all the steps in the acquisition of mining property, except in prospecting, the shaft mine appeared to be discriminated against. The claiming of a field was made extremely costly and involved greater risk for the shaft mine. The concession and the related measuring and size of the field made shaft mining operations less profitable. The law of 1821, which was to provide large fields benefitting the shaft mine, was not actually applied until the 1840's.

The case of the Schölerpad mining company provides a concrete example of control of production by the mining administration. It also provides an insight into the economic reasoning behind the Direktionsprinzip. From 1816 to 1826, Schölerpad was continually refused permission to start a shaft mining operation, although it possessed the required legal concession for the field itself. The mining office gave the following explanation. First, there already existed a shaft mine, Sälzer-Neuak, in this area. It supplied a sufficient output for the coal demand in the same area. Second, although Schölerpad had the field concession, this did not entitle it to demand permission for the operation of the mine. The law of 1807, granting Gewerbefreiheit, freedom to practice one's occupation, explicitly did not extend to mining. Third, if Schölerpad were given permission to mine, Sälzer-Neuak would suffer disastrously. It was located one-half hour farther from the Cleve coal market than Schölerpad. Consequently, Sälzer-Neuak would lose a large part of its coal market and would be unable to cover its cost and debts previously accrued. Fourth, the public would not benefit from the resulting competition, since a price increase would inevitably follow because both mines would attempt to cover their costs.[2]

For ten years Schölerpad fought back. It presented its case to the Oberpräsident of Westphalia, von Vincke, and to the Geheimen Staatsminister vom Stein, and finally to the king himself. To no avail, as the cabinet order, specifically directed to the Schölerpad mining company, made clear. This order of 1826 again reaffirmed the mining administration's sole right to grant a Betriebserlaubnis.[3]

Nevertheless, the Schölerpad case did achieve the result that the king himself in this order let it be known to the mining administration that "one could not grant Sälzer-Neuak automatically a monopoly, by not allowing the operations of the neighboring shaft mines."[4] This had the consequence that in the future the mining administration was more lenient and granted Betriebserlaubnis if the company involved was persistent in its demand. There were other cases where permission to start mining was granted only after the output of the mine was strictly limited and a definite market provided, so as not to compete with existing mines.

In the case of the mine Helene-Gertrud the mine owner did not receive permission to start operation until he promised to use the coal output only for his own brewery.[5] The mine Timmerbeil was allowed to reopen its operation in 1828, only after it had assured the mining office that the coal would be sold solely to the iron works at Witten.[6]

Franz Haniel was the first entrepreneur to risk the construction of shafts penetrating the water-bearing layer. In the construction of his first shaft, Franz, (1832-1834) he failed. Between 1835 and 1838 he successfully constructed the shaft Kronprinz at a cost of 60,000 thalers. The mining office at Essen attempted to discourage him from bringing this shaft into operation.

Haniel, in seeking permission to start operation, gave as reasons that coal prices were high and that the demand of the ironworks, Guthoffnungshütte, had not been supplied. The mining office on the other hand argued that the price increase was only temporary because of the recent export of coal to Holland. Further, the office stated that the ironworks certainly did not suffer from lack of coal supply. Haniel was an owner of these ironworks. He disagreed, giving numerous examples where the ironworks were not supplied with the quantity and quality of coal they needed. Finally permission was granted, but the mining office demanded immediate payment of the dues of 3,660 thalers for measuring the fields.[7]

The shaft mine Vereinigte Helen und Amalie had to stop shaft construction in 1842. The mining office claimed that the mine had not received police permission to operate the steam engine used for lifting, since the required technical plan of the steam engine could not be located at the mining office.[8]

In 1842 the Oberbergamt was trying to hinder the construction of a shaft at the mine Heinrich. It argued that the coal market was insufficient. Furthermore, it attempted to block the operation by requiring a vote of the shareholders. But the shareholders were mostly coal dealers and they apparently saw the market condition in a more favorable light and voted for the operation. Despite this vote, the Oberbergamt still prohibited the operation. Only after continuous demands from the mine owners did the office finally agree to grant permission in December, 1847.[9] The shaft construction was immediately started and production began in 1852. (See Table 4.4)

In general, the mining office granted permission to start operation only after persistent demand by the mine owners; during the 1840's the mining office gradually relaxed its control over the structure of the industry by allowing large shaft mines to enter the industry. Nevertheless, in 1843 Heintzmann, the leading mining official of the Oberbergamt, reluctantly admitted "that a decrease in output for the industry is probably no longer possible, since every mine is allowed to begin operation."[10]

The coal industry's structure was compellingly determined by geology. Obviously a coal mine has to be located where nature put the coal, regardless of where the consumer or the operator would like it. Moreover, geology did not make coal measures uniform either in quality or accessibility.

Variation in quality, location, ease of working the coal wherever it may be found and the way it can be combined in industrial partnership with other resources--all influenced the industry's structure. The important characteristics of the geology of coal reserves are the quality of the coal, the thickness of the seams, and their depth below surface. Therefore, the kind of underground operation required, shaft, slope, or drift mining, is important.

There are two major characteristics of the geological nature of the Ruhr coal field. The first is that the coal is found at progressively increasing depth from the south to the north of the field. In addition, the depth increases more rapidly in the western part, the Essen-Werden area, than in the eastern part. The second characteristic is that the coal eventually lies under the water-bearing secondary rock. This was the case of the _Fettkohle_, or coking coal.

The effect of these geological features of the coal was that as mining moved northwards the initial cost of mining increased. Furthermore, the mining activity thereby also moved away from the traditional transportation route, the Ruhr River. As long as mining was in the southern fringe of the coal region, the Ruhr River and its side valleys, access to the coal was fairly easy and the cost of extraction was relatively small and constant. Open-pit, slope, or tunnel mining were the typical operations. By the beginning of the nineteenth century many areas around the

Ruhr Valley were exhausted. This was increasingly the case of mining in the Essen-Werden area, where shaft mining operations were begun relatively early.

The first shaft mines in the Essen-Werden area did not have to penetrate the water-bearing layer, although they did have to cope with water problems. But the continuous expansion of production required the deep shaft mine and the severe water problem demanded a new technology. The steam engine, first used for water pumping but later also for lifting the coal, became the necessary companion of the shaft mine. Finally, the shaft mine with its increased output and demand for larger and more distant markets upset the small scale mining nurtured by the Direktionsprinzip.

The water-pumping and coal-lifting steam engine transformed mining technique, making deep shaft mining possible. In 1803, Franz Dinnendahl produced the first Ruhr made steam engine and delivered it to the mine Vollmond. The mine found it difficult to pay the lump sum of 2,400 thalers; it had to get a loan from the miner's fund. Before 1815, there were about a dozen small steam engines pumping water in the Ruhr mining region. As tables 4.0 and 4.1 show, nine of these were in the Essen-Werden region where all of the early shaft mining was taking place. Soon most of these turned out too weak for the handling of the water. Already in the 1820's stronger ones became necessary.[11] For instance, the mine Kunstwerk replaced its 1813 engine with more powerful one in 1822 and again in 1827. The same was true for the mines Sellerbeck and Gewalt.

Furthermore, as tables 4.0 and 4.1 show, early shaft mining was costly and risky. By the 1820's the more powerful and complex steam engines already were priced over 20,000 thalers. Financial success was uncertain; some shaft mines made large profits, others had large losses and had to shut down. Finally, it is noteworthy that while most of the early steam engines were produced by the Dinnendahl family and Franz Harkort within the Ruhr region, engines by Cockerill and a French firm were already entering the region.

The number of steam engines steadily increased. In 1834 there were 50; in 1843 the 48 shaft mines had 95 engines.[12] By 1850 there were 123 and some of these were already used for lifting coal. With the introduction of the steam engine the water problem of the shaft mine had become manageable.

Shaft mining changed the structure of the industry by increasing the size of the individual operation. The typical mining operation throughout the first half of the century had an average output between 1,000 and 5,000 tons. Even in 1850 158 mines, or 60 per cent of the total of 198, had an average output of only 1,500 tons. In the same year the largest mine, Gewalt, already produced 84,000 tons. Still, by modern standards, the concentration of production was not impressive in 1850. The largest 10 mines shared 40 per cent of the total output and the largest 40 mines 87 per cent (see Tables 4.2 and 4.3). Nevertheless, by the standards of the Direktionsprinzip, nurturing small mines, this was an impressive change.

It is significant that despite the continuous opposition by the mining administration, nearly all of the 40 mines were engaged in some kind of shaft mining operation. Furthermore, it is significant that only 7 from the 40 largest mines had penetrated the water bearing secondary rock and were producing in 1850. But these seven mines marked the beginning of a new era in Ruhr coal mining.

The following generalizations can be made about the first water penetrating shaft mine after 1830. First, this type of mine developed despite the opposition of the mining administration. Second, it required money, influence, and the willingness to take risks. Third, it marked the beginning of truly large scale mining in the Ruhr.

The opposition to shaft mining has already been discussed. In the case of the new type of operation a small group of profit seeking but also far-sighted entrepreneurs took the initiative and financial risk to overcome not only the physical obstacles but also the administration's attitude. This could only be accomplished with money and influence. As Table 4.4 indicates, the same people for the most part engaged in this new venture. Haniel, Stinnes, Harkort, Waldthausen, and Butenberg had money and were making more money in the process. Von Eicken, Honigmann, and Von Vincke had connections and influence. Von Eicken and Honigmann were retired mining officials and Von Vincke of course was the president of Westphalia.

Already in 1850, after only a few years of operation, all seven mines were among the largest in the Ruhr region. The two Stinnes mines, Victoria Mathias and Graf Beust, were the third and fourth largest mines in the Ruhr region. The first and second place mines were very old mines; Vereinsigte Hagenbeck was the oldest officially known mine in the Ruhr region, dating from the sixteenth century. The mines Helene und Amalie, Franziska Tiefbau, Vereinigte Präsident, Carolus Magnus, and Wolfsbank (see Table 4.4) were already by 1850 respectively the 8th, 12th, 16th, 25th, and 32nd largest mines in the region.

B. Marketing and Market Controls

1. Marketing

In the eighteenth century the Prussian state and particularly Frederick the Great were trying to foster the growth of the coal industry in the western territories. The state encouraged the inflow of trained miners, especially from Saxony, Thüringia, and the Luttick mining region. By 1755 the coal mining employment in the County of Mark was 688 workers, of whom 57 were from outside the Mark. By the 1770's a sizeable permanent and professional mining community

TABLE 4.0

EARLY STEAM ENGINES AND SHAFT MINES IN THE ESSEN REGION

Year Starting Shaft Mine Operation with Steam Engine	Name of Mine	Cylinder Size (Zoll)	Shaft Depth (Meter)	Operation to year	Evaluation Profit or <u>Zubusse</u>
1805	Nottekamp-Bank	30	62	1827	13000 Thaler <u>Zubusse</u>
1808	Rosendelle	36	47	1817	<u>Zubusse</u>
1810	Gewalt	26	13	1814	2500 Thaler Profits
1812	Wiesche	40	44	--	Profits
1813	Sonnenschein (later Kunstwerk)	28	35	1820	Large Profits
1813	Caroline	44	45	1817	Large Profits
1814	Wasserschneppe	28	67	1821	36000 Thaler <u>Zubusse</u>
1814	Kleffklappen	26	24	1819	<u>Zubusse</u>
1814	Sellerbeck	26	96	--	<u>Zubusse</u>

Source: Reuss, "Mitteilungen...", <u>Festschrift</u>, pp. 63-65.

TABLE 4.1

STEAM ENGINES IN THE ESSEN REGION: 1821-27

Year of Installation of Steam Engine	Name of Mine	Cylinder Size (Zoll)	Shaft Depth (meter)	Price of Steam Engine (Thaler)	Name of Producer
1821	Gewalt	34	104	--	Cockerill
1822	Kunstwerk	30	-	--	F. Dinnendahl
1827	Kunstwerk	44	-	14530	J. Dinnendahl
1824	Sälzer & Neuak	48	100	23700	F. Harkort
1827	Sellerbeck	50	88	21000	Engbert Ruleaux, & Dobbs

Source: Reuss, "Mitteilungen...", Festschrift, pp. 63-65.

TABLE 4.2

SHARE OF LARGEST MINES OF 1850 OUTPUT

Largest Mines	Average Output (tons)	Percentage of Production
Largest 10	60,000	40.0
Largest 20	54,000	59.0
Largest 30	41,000	75.0
Largest 40	37,000	87.0
Remaining 158	1,500	13.0
Total 198	8,400	100.0

Source: Die Entwicklung des niederrheinisch-westfälischen, Steinkohlenbergbaues..., Vol. X, p. 53.

TABLE 4.3

PRODUCTION OF THE TEN LARGEST SHAFT MINES IN 1850

Name	Production (tons)	Employment
1. Gewalt	84,000	356
2. Vereinigte Hagenbeck	76,000	394
3. Victoria Mathias	62,000	333
4. Graf Beust	59,000	334
5. Vereinigte Sellerbeck	59,000	311
6. Vereinigte Nachtigall und Aufgottgowagt	57,000	217
7. Vereinigte Sälzer und Neuak	56,000	332
8. Helene & Amalie	55,000	351
9. Friederich Wilhelm	49,000	-
10. Freie Vogel und Unverhofft	46,000	-

Source: Entwicklung..., Vol. X, p. 53.

TABLE 4.4

FIRST SHAFTS PENETRATING THE WATER-BEARING ROCK (MERGELDECKE)

Mine or Shaft*	Shaft Construction Started	Uncovering Coal to 'eyesight'	Production Started	Major Shareholders
Franz[1]	Fall 1832	March 1834	failed	Franz Haniel
Kronprinz	July 1835	Aug. 1838	May 1838	Franz Haniel
(Müller)	Oct. 1834	----	Jan. 1839	Von Eicken
Wolfsbank[2]	Apr. 1838	Dec. 1840	1844	Brockhoff and Von Covy
(Herkules)	Dec. 1838	Dec. 1840	failed	Franz Haniel
Graf Beust	May 1840	March 1841	Jan. 1842	Mathias Stinnes
Präsident	Summer 1840	Jan. 1841	Fall 1844	Ludwig Von Vincke and Friederich Harkort
Joachim	Summer 1840	Dec. 1841	gave up	J. W. Butenberg
Victoria Mathias	1840	-	1845	Mathias Stinnes
Franziska	1837	-	after 1845	--
Helene und Amalie	1842	-	1846	F. W. Waldthausen
Carolus Magnus[3]	-	-	1846	Earnst Honigmann
Engelsburg	1846/47	-	end 1850	--
Heinrich[4]	1847	-	1852	Waldthausen, Haniel, Butenberg
Zollverein	1848	-	1852	Franz Haniel
Hannibal[5]	1848	-	1854	J. H. Von Hüllen

Source: Hans Spethmann, _Die ersten Mergelzechen_,... and Gebhardt, _Die Geshichte des Ruhrbergbaues_ ... and Bierhaus,.... p. 39

* Name of shaft in parenthesis

1. Production was discontinued in 1842

Table 4.4 - continued

2. Wolfsbank was bought in 1847 by the newly founded Kölner-Bergwerksverein (Cologne Joint-stock Mining Company)

3. Ernst Hongimann, a mining official, had received the concession in 1841 and sold the mine to the Stinnes family in 1847

4. Was bought by the Cologne Joint-stock Company

5. Von Hüllen and his partner H. Heyparty had already drilled for the coal in 1846-48 before constructing the shaft in 1848. The result of drilling was not accepted as sufficient evidence to receive a field concession until after 1853.

had come into existence.[13]

Under Frederick the Great the state forcefully intervened to provide a market for coal. The state and its leading officials tried in every way to persuade, coerce, or enforce the use of this new fuel. In the local iron and coal industries persuasion was tried. Frederick the Great promised a reward of 50 thaler for anyone who could discover how the Osemund ironworks in the Mark could use coal instead of wood. Only in 1790 was the first attempt made with coal at the Antonihütte, but it failed. The local Mark brewing industry was ordered to use coal or pay a fine. The bakeries were asked to use coal since the king had proven in his royal bakery in Berlin, baking army bread, that coal could be used. Mining officials in the Mark tried to encourage the use of coal in the households for daily cooking. Already in the edicts of 1766 and 1769, Frederick the Great ordered the use of Mark coal rather than foreign coal in the western territories.[14]

The Ruhr region as a coal market was rather small. Despite a gradual population growth and increasing industrial activities, the towns in the Ruhr region remained relatively small compared to the older cities on the Rhine. Even by the end of the eighteenth century none of the Ruhr region's towns were over 5,000 and most of them were between 800 and 3,000.[15] Consequently if the industries, especially the coal industry, were to develop they depended on the expansion of the market outside the Ruhr. This in turn depended on transportation development, already discussed in chapter III.

The opening of the Ruhr River to navigation and the linkage with the Rhine resulted in economic unification of Prussia's most important western territories, the coal and agriculturally productive Mark and the more industrialized Duchy of Cleve, as well as the smaller territories of Meurs and Geldern, on the Rhine. In general, in the second half of the eighteenth century free navigation of the Ruhr was a prerequisite for the industrialization of the Ruhr region.

The opening of the Ruhr had a direct impact on coal mining. The mine owners searched for convenient places near the river. But even more significant in the late 1780's and 1790's the number of claims and concessions increased to the point where nearly all the conveniently located coal layers near the river were claimed. The expansion and concentration of coal mines in the area near the river gave this area "for the first time a truly industrial mining character."[16]

In 1786, the year of Frederick's death, Minister Von Heinitz, after an inspection tour of the Mark, was able to report:

> The coal mines of the County of Mark are the most important in Your Majesty's dominions. Last year the output of coal was 1,707,461 scheffels... In 1727 the output of coal amounted to only 467,875 scheffels. These coal mines supply the industrial

> undertakings of the County of Mark and the provinces of Cleve and Gelders with all the fuel that they require. There is also a very considerable export of coal from the County of Mark to the Duchy of Berg and the Bishopric of Münster and some of the territories of the Bishopric of Cologne. Even Holland imports some coal from the County of Mark. The opening of the navigation of the River Ruhr by the construction of locks and improvement in road maintenance had facilitated the transport of coal from the hilly districts of the county of Mark...[17]

As this report indicates, the opening of navigation on the Ruhr, although the 17 locks had not been completed until 1790/92, had already an impact on the coal industry and its market.

The opening of the entire 72 kilometers of the navigable part of the Ruhr caused a production increase of 45,000 tons or nearly 35 per cent from 1791.[18] Coal could now be shipped from the interior of the Mark region to the Rhine, particularly the Prussian Duchy of Cleve. The physical market area had been expanded. During the first half of the nineteenth century the marketing and consumption pattern had changed drastically. By 1851 the consumption pattern of Ruhr coal was as follows: More than 47 per cent was industrial and household consumption within the Ruhr region. Consumption by iron and steel works within the region was 7 per cent. Consumption by mines themselves was also 7 per cent and consumption by the railroads had reached 3.5 per cent. Finally shipment from the Ruhrort-Duisburg ports was about 35 per cent, constituting direct export of coal.[19]

The rate of population increase in the Ruhr region itself indicated a healthy growth during the first half of the nineteenth century. In 1818 the official administrative district of Dortmund contained 304,000, the district of Bochum nearly 300,000, and the district of Hamm, 310,000. By 1852 the Dortmund, Bochum, and Hamm districts had increased to 700,000, 600,000, and 470,000 respectively. In 1818 Mühlheim was still the largest city in the Ruhr mining region, even ahead of Dortmund with 4,700. By 1832, Dortmund moved ahead with 7,000 while Mühlheim had only 5,000. By 1855 Dortmund, with a population of 22,000, became the administrative center for the entire Ruhr region.[20]

Many of the smaller cities on the Ruhr River owed their life to the river and the shipment of coal. This was particularly the case for Mühlheim. Duisburg, although it had access to the Ruhr River, was a Rhine port. During the first half of the century the Ruhrort port facilities were constantly improved and expanded, and by the 1850's it had become the largest inland port in Western Europe. In general, the real growth of most Ruhr cities came in the 1850's and 1860's, but between 1820 and 1850, the Ruhr region's population more than doubled.

The iron and steel industry had become a major consumer of coal. The infant iron industry got its real start only after local blackband deposits were discovered and coke could be produced. A small iron industry already existed in the eighteenth century. It was mainly based on a local supply of _Rasenstein_, an inferior grade of iron ore found in some marshy areas of river valleys. The production of iron ore between 1815 and 1840 ranged between 2,000 and 6,000 tons. Not until blackband was discovered did iron ore production become significant.[21]

Blackband was found in the 1830's near Bochum while mining for coal, but no one paid any attention. In the mid-1840's, Schreiber, a state mining official of Hesse, passing through the Ruhr region saw the iron ore piled up. He had just returned from an inspection tour in Scotland. There blackband, a few years earlier, had provided the push for the development of the iron industry. Schreiber immediately made the owners of the iron works, Hermannhütte, among them Gustav Mevissen, aware of the blackband. As the Table 4.5 shows, in the late 1840's a great search for blackband began and field concessions dramatically increased in 6 years from none in 1847 to 71 in 1852. By 1850 the production of iron ore had already reached nearly 12,000 tons.

Coke production also had to wait until shaft mining moved northwards. _Fettkohle_, the coal suitable for coking, was only found under the layer of water-bearing secondary rocks. In 1835/36 only Sälzer-Neuak made some primitive coke. By 1841 coke production was 12,000 tons, by 1846, 32,000 tons, and by 1850 it had increased to 40,000 tons.[22]

Fritz Harkort introduced the puddling process in the

Ruhr region in the 1830's. With the help of an English puddle master he established the first puddle-rolling works in the Ruhr region at Wetter. It was from here that the puddling process gradually spread. According to contemporaries, the puddling process made it possible to produce in three months the same amount of iron that was formerly produced in three years.[23]

In the 1830's Fritz Harkort attempted unsuccessfully to introduce improvements in smelting. Here coke became crucial. The early attempts failed because usable coke could not be produced. In 1837 the pig iron production of the smelting furnaces in the Ruhr region was 5,300 tons, produced by six blast furnaces, all using charcoal. Even by 1841 the nine blast furnaces producing 8,000 tons used charcoal exclusively. In the following decade coke was introduced and by 1851 the fifteen blast furnaces produced nearly 12,000 tons pig iron, of which more than 40 per cent was produced with coke.[24]

The export market played a crucial role for the development of the industry. In the first half of the nineteenth century, with improved navigation on the Rhine, particularly Holland became in the 1830's an important market for Ruhr coal. But the biggest obstacle to coal export was Holland's tariff on Ruhr coal. Before 1830 the export was rather limited. Not until 1830, when Belgium separated itself from Holland and Holland reduced its tariff, did coal exports become significant. However, after relations between Belgium and Holland were normalized, Holland again raised its tariffs.

The importance of Holland's coal market to the Ruhr region can be seen in the five years, 1831-1835. During these years the tariff on Ruhr coal was abolished. According to the contemporary, Carnall, the abolition of the tariff resulted in "an upswing of coal prices, profits and general mining activity which had never been before."[25] In the Essen-Werden mining area, employment increased by 84 miners within a month; in the Bochum area 15 mines increased their total employment by 110. By the end of the year 1831, the export of coal to Holland was 154,000 tons, or nearly one fourth of the total Ruhr production.

The effect on the coal production from the access to Holland's market is indicated in Table 4.6. It had two major consequences: (1) the immediate effect of a boom, 1831-35, (2) affecting the structure of the industry by encouraging shaft mining. The boom is evidenced by a sudden production increase of 115,000 tons by the end of 1831. The number of mines increased from 1830 to 1834 by 44 and total employment increased by 1400 from 1830 to 1835. While there was no dramatic price fluctuation in the average price for the entire Ruhr even this price gradually increased from 5.72 R.M. to 6.07 R.M. per ton.

This boom encouraged the more costly shaft mining operations. Many of the new mines during the 1830's were shaft mines. The boom had the effect of depleting the more easily accessible coal and making a more rapid northward movement of mining necessary. Particularly in the Essen area, the deep shaft mines, penetrating for the first time the water-bearing secondary rock layer (<u>Mergeldecke</u>) were begun in the late 1830's.

In 1835, when trade between Belgium and Holland resumed, Holland raised its tariff on Ruhr coal to nearly three fifths of the value of the coal. As Table 4.6 indicates, this had the effect not only of driving out 26 mines in the same year, but also of decreasing production from 784,000 tons to 745,000 tons in 1836. Since this decrease in mines is not reflected in employment (instead employment increased sharply by 900) one must assume that consolidation of mining operations was taking place: average employment per mine increased from 25 to 31, the largest increase during the entire 1830's. In addition, mines which had begun their shaft construction during the boom period began altering the size of the operation. Both the consolidation of existing mines and the entrance of new shaft mines caused once more a large increase of production during 1837 and 1838. In general, between 1830 and 1840 employment doubled, production nearly doubled, the size of the average mining operation increased: output per mine by 83 per cent, and employment by nearly 60 per cent.

Nevertheless the Ruhr coal was gradually driven from the Holland market. Of course the high tariff was one reason. Another reason was the costly Ruhr River transport, discussed in Chapter III and the price and tax policy of the <u>Direktionsprinzip</u> creating artificially high prices. Consequently English and Belgian competition drove Ruhr coal from the market. According to the leading mining entrepreneur and coal dealer, Franz Haniel, English coal in 1837 was far cheaper than Ruhr coal.[26]

Despite Prussia's effort after 1839 to give encouragement by granting tariff rebates to export transporters on their return trips, coal exports continually declined. By 1839, coal exports to Holland were at the same level as in 1832, but total production had increased. By 1844 exports had decreased to 90,000 tons. Again the Prussian government gave additional tariff rebates. Therefore, by 1850, exports amounted to 150,000. This was, however, a small percentage of the total Ruhr production of nearly 1.7 million tons.[27]

2. Market Controls: Price System

During the first half of the nineteenth century, the control and regulation of sales through maintenance of a fixed price system failed. The mining administration no longer possessed an appropriate view of the market. The objective of the regulation of sales had been to regulate competition by providing each firm with an adequate market share. But the expansion of the market and production made the regulation of sales and competition increasingly more

TABLE 4.5

MINING ACTIVITY IN THE IRON ORE (BLACKBAND) INDUSTRY: 1847-1852

	Searching Warrants		Claims		Concessions	
Year	Applied	Granted	Applied	Granted	Applied	Granted
1847	220	90	50	31	2	0
1848	99	65	119	25	10	6
1849	112	67	33	15	10	13
1850	400	322	295	202	20	11
1851	628	509	426	228	60	51
1852	901	765	499	262	82	71

Source: Entwicklungen..., vol. XII, p. 13.

TABLE 4.6

COAL PRODUCTION: 1830-1840

Year	Number Mines	Total Employment	Total Production (1000 tons)	Output Per Mine (tons)	Output Per Worker (tons)	Average Price Per Ton (R.M.)	Workers Per Mine
1830	172	4500	511	2970	114	5.72	26
1831	180	4700	626	3480	133	5.85	26
1832	194	5200	675	3480	130	5.81	27
1833	211	5400	761	3610	141	5.95	26
1834	216	5400	767	3550	142	6.03	25
1835	190	5900	784	4130	133	6.07	31
1836	199	6800	745	3740	110	5.72	34
1837	208	7500	868	4170	116	5.62	36
1838	203	8100	996	4770	123	6.28	39
1839	213	8300	1007	4730	121	6.61	39
1840	221	9000	990	4480	110	6.27	41

Source: Reuss, "Mitteilungen...", Festschrift, p. 82.

difficult. Indeed, it appears that after the 1820's the major reason for maintaining sales controls and enforcing the government established prices was to collect the mining taxes, computed on the basis of these prices.

Carnall, a member of the mining law reform commission in the late 1840's, explained why the price and tax policy was difficult to administer:

> "As long as there were only a small number of mines and the coal had been sent to the same market such equalization through differential taxes and prices may have been possible. Now this is no longer true and has not been possible for some time. And it is today no longer strictly enforced but rather the taxes and prices are determined in conference with mine owners for the coming year. Because of existing conditions, the competition among the mines and the total change in the market through the railroads, one can no longer apply the terms of the old law."[28]

In general, after the 1820's, market expansion made the regulation and control of sales and the establishment of differential prices extremely difficult if not impossible. The improved and expanded transportation facilities and the variations in distance involved made it difficult to account for the transport cost in the individual mine price.

The price-tax system was maintained until 1851. The application of this system was less strict by the 1830's. The method of determining the price was changed. At a general meeting with the mine owners the mining office established the prices for the coming year. Legally, the mining office was still solely responsible for the establishment of prices and the mine owners were there only in an advisory capacity.

This practice could have led to the establishment of a true market price. After all, by the 1830's, the mass of the mine owners were better informed about the actual market condition than a rather small group of mining officials, however competent. The mining office itself admitted that it could not establish a price for each mine and still fulfill its other administrative functions. It had to accept the advice of the mine owners.

The yearly price meeting did not lead to the establishment of true market prices, although it reflected these. Frequently the mine owners established an independent price among themselves. This tendency had been inadvertently encouraged by the finance ministry's directive to the *Oberbergamt*. The directive provided a basic price guideline: "as long as there is no evidence for a change in prices, either an increase or decrease, the assumption should be that the old price was to remain."[29]

For a price change, the mining office was increasingly dependent for information on the mine owners who of course would protect their own interests. The result was that at the meeting the majority decided the price. The problem with this procedure was that the majority were the owners of small mines with a local market, while a minority possessing the larger shaft mines and exporting the coal faced a different market and price. A mining report of the price meeting in 1837 reflected the above problem:

> In this year's meeting to determine the price for 1838, the overwhelming tendency was to favor a slight but imperceptable price increase for the distant consumer. There was opposition from the minority, owning the best mines and enjoying the larger profits. However, the decision wisely reflected the majority interest.[30]

These yearly meetings began to create the opinion that the price fixing was a yearly contract between the mining administration and mine owners. Furthermore, the established price was used by the mine owners, not necessarily as a true price, but as a tax guideline. This in turn encouraged the practice by mine owners of selling at other than the established price, particularly higher than the government price, to gain a tax benefit. Consequently, at the yearly price meeting, there was in general a tendency to press for a lower price level so as to lower the tax base.

The yearly meetings did not establish a uniform coal price for all mining districts, but they did influence the price level in general. Separate prices were established for various groups of mines depending on their production and marketing facilities. Furthermore, the mining office still attempted to set prices for individual mines, if it felt it was absolutely necessary.

In general, the individual or group mine prices did not vary greatly from the general price level of the individual district mining area. But the general average price of the Essen _Bergamt_ region was different from that of the Bochum

Bergamt region. The major reasons for the difference were the higher quality of the Essen coal and its access to export markets in the Rhine Valley and Holland.

For instance, in the 1830's a mine or group of mines whose owners were also coal dealers and transporters would have different market opportunities and actual market prices. The price of coal at Cologne was probably higher than the price in Holland where the Essen coal had to compete with Belgian and English coal. In addition, the price in the Ruhr region itself, although it most closely reflected the established price, could also be different. Therefore all three prices could in any given year be in variance from the government price.

Mines which exported coal were certainly more dependent on the actual export market price which could fluctuate violently even within a given year. Yet, the annually established price was decided by a majority who did not necessarily face these market conditions but rather had a fairly stable regional market. The following can be said about the generally established price. It did reflect the actual market price in the long run but reflected less the short run market fluctuation. While it served for mines whose market was local and within the region as the actual market price it did not accurately reflect or directly react to the export price. In general, after 1830 the price served as guide below which it was not profitable to sell because taxes would be computed from the established price. Finally, after 1830 the price tended to be below the actual market price because it was used for tax computation, and taxes were felt as the real burden.

The mining administration also attempted to regulate the procedure of coal sales. Sales on credit were specifically prohibited when they involved land transport. In the case of sales which took the Ruhr River route, only after 1827 were the mine owners allowed to give six months credit. The major reasons were the seasonal water problems of the rivers Ruhr and Rhine and the distance to the export market. But this was only possible if two thirds of the shareholders agreed.[31]

The physical division of the output among the shareholders was practiced by mines located near the Ruhr River in the Essen-Werden area. This method was especially favored by mine owners who were also coal dealers and transporters. This was to the disadvantage of those shareholders who were not coal dealers because they were forced to sell their coal to the latter. They, of course, complained. The practice of dividing the actual coal was officially prohibited in 1827 by the _Oberbergamt_.[32]

The mining office had no jurisdiction if the entire output of coal was bought by a coal dealer who also happened to be a mine owner. This was legal as long as the coal was bought at the established government price and the majority of the shareholders had agreed. Consequently, after 1827, the coal dealers and transporters attempted to get control over the mines in which they owned shares. Under these conditions the mining administration no longer had control of the sales of coal at the market, nor of the actual market price. This practice obviously tended toward concentration of ownership.

In 1839 the Finance Ministry instructed that in the future the coal sales of a mine should be undertaken by an elected representative who was not a coal dealer.[33] This instruction did not really affect the larger shaft mines, which were owned either by one person or a few shareholders with common interests and had no intention of obeying the vague instructions. Furthermore, the ministry had only acted upon the complaints from a vocal mine owner, L. Miller, who felt he had been cheated by fellow shareholders and mine owners. The ministry had no real intention of restricting further the sale of coal.

In general, during the 1830's and 1840's, particularly in the Essen-Werden region near the Ruhr River, the resale of coal was practiced by most mines. Furthermore, by the 1840's, after the introduction of the railroads, the mining administration not only had little knowledge of the actual market price of coal at the numerous and distant consumption points, but could no longer effectively control the actual sales on the market.

L. Miller, a mine owner, gives one of the most interesting accounts of the malpractices of the 1830's and 1840's. Since he felt he had been cheated, he was angered by his fellow mine owners and proceeded to point out the illegal practices of his colleagues. He collected documented evidence and wrote letters to the mining offices asking them to stop the insults to the law.[34] By his own admission, he was angry because some of his fellow mine owners "had several times cleverly cheated him out of his share of the profits."[35] But his appeal for "justice" and "honesty in adherence to the law" apparently fell on deaf ears.[36]

Miller's account describes more concretely how prices were set at the early meetings. A typical meeting was held on November 6, 1839, in Essen. "The committee representing the various mining companies again applied for prices far below the current market price as it had always done in previous years."[37] Miller wrote, "I protested the acceptance of such prices."[38] This meant that lower prices were set to lower the taxes, the tenth, and many mines sold at higher ones. He gave an example. The market price in 1839 for good brick coal was 6 1/2 to 7 1/2 Silbergroschen per scheffel, but at the meeting the government price was set only at 5 Silbergroschen per scheffel for the next year. This price would also be listed in the official royal Prussian records at Berlin, "insulting his majesty, the king." "Actually, at first, they had the audacity to ask for a lower price than the five Silbergroschen and were only grudgingly giving in to the higher one."[39]

Miller reported how mine owners were circumventing the law and selling at other than government established prices. The mine Langenbram was the most notorious of these "questionable mines." Others also had admitted that they were not selling at the established price. The representatives of Langenbram had already sold their 1839 production at the lower price set in the previous 1838 meeting to its own shareholders who also happened to be coal dealers: Georg Stinnes, Neustein (who was also a representative at the meeting), the Brothers Schwengers (brother-in-laws to Neustein), and the Brothers Maurits and Smith.[40] All in turn sold the coal at the higher current market price of 1839. What made Miller even more angry was that "everybody knew it." How could the mining office not know it? "Especially since several representatives immediately after the meeting publicly thanked the king's officials for the favorable treatment they had been receiving."[41]

Miller also documented monopolistic price fixing agreements. Most of the mines set a sales price by common verbal or even written agreements. The representatives of the mine, Vereinigte Kapellenbank, had by written agreement with other mines established a fixed price to which every participant had to adhere or forfeit a fine of 100 thalers.[42] Among most of the mines there were also verbal agreements on prices, as Miller himself knew by his earlier participation. "This price fixing also took place immediately after this year's meeting."[43]

The same kind of price agreements had also been taking place in 1838. In the 1838 price meeting, the mines in the Essen-Werden area had already formed a price agreement among themselves. The mines publicly announced that they would not sell at the official price.

In the same year, 22 mines in the Essen-Werden area created a joint-stock company with a capital of 8,000 thalers to sell their coal in common. Three days prior to the meeting, they had already agreed to sell their coal at a commonly established price. Due to disagreement among the mine owners, however, the company failed.[44]

The mine "Vereinigte Kapellenbank," the rebellious leader, in 1838 even defended before the mining office the discrepancies between the tax-price and the true market price. It argued that this difference, was "a repayment for the effort and risk incurred in the sale."[45] It is noteworthy that the emphasis is on the effort and risk in the marketing rather than in the production of coal.

Another way of circumventing the control of sales, in this case directly at the mine, was the practice of giving discounts to favored customers. The Direktionsprinzip, especially in the eighteenth century, had attempted to establish a uniform measuring system. The Prussian

"Berliner Scheffel," with a content of 3.72 Prussian cubic inches, became after 1816 the uniform measure of coal. The "Scheffel" was a wooden basket. But even a uniform measuring system did not guarantee the mining office exact control over the output and sales of a mine. Even if the price was enforceable at the mine, the discount to large quantity buyers could be given in form of heaping more on the individual "Scheffel." The mining office of course wanted an evenly filled basket.

Particularly in the 1830's, with the beginning of shaft mining, the practice of discounts via additional coal spread. The shaft mines wanted to guarantee themselves a larger market share. For instance, it was reported that the Prince Wilhelm coal railway in 1836 had bought 1,000 Scheffel but actually received 119 extra Scheffel or more than 11 per cent. In 1836 a mining report mentioned that the mine St. Morits had measured out at least 20 per cent more coal.

This form of discount was repeatedly declared to be unlawful, particularly since it effectively lowered the taxable income, but in general the mining offices could not police it. For instance, the coal freight cars of the Cologne-Minden railway were supposed to hold 80 Scheffel but in reality they were always loaded with 105 Scheffel to be sold as 80 Scheffel. The mines "Helene and Amalia," "Wolfgang," and "Steingatt" gave on the average a 30 per cent discount. In the southern area of the Mark, where most of the coal was transported by land, the mines even more than 30 per cent extra coal.[46]

Despite all these practices for circumventing the government controls, the mining officials of the Oberbergamt still attempted to enforce the sale of coal at governmental prices. In general the mining office did not object to and even directly permitted the sale of coal lower than the government price as long as the taxes were computed on the basis of the government price. In some years the uncertain Ruhr River transport made it necessary to sell the coal locally at lower prices. The same was true in the case of land transport. Of course, these sales below the established tax-price did not affect the state tax income.

The practice by mine owners of giving discounts to certain consumers of large quantities was prohibited by the mining office. As long as taxes were paid on the non-discounted price, it was permitted. As late as 1851 the mining office at Bochum threatened its district officials with a fine if they did not report the cases where the discount price rather than the government price was used as the basis for tax computation.

Selling higher than the government price, and not using this higher price for the basis of taxes, was of course prohibited. Occasionally the mining office did allow a higher price than the established one when transportation, particularly the Ruhr River transport, made it necessary. In this case, though, the higher price was to be used for the tax. If the mining office did have evidence that a mine sold higher than the established price and the higher price was not used for tax computation, the mining office did step in. In 1844, the *Bergamt* asked the *Oberbergamt* to penalize the mine Langenbrahm for a price discrepancy of this sort.[47]

The greatest opposition came when the mining office raised the tax-price for certain mines after the general price had been established for the year. The mining office of course had the legal right to do so. It pointed out that according to the 1766 codification, Article LXXIII, paragraph II, the *tenth* was to be paid on the basis of the actual value of the coal. This of course was detrimental to those mines which had already made a contract with their consumers at a fixed price.

Most of the high officials at Berlin and high political officials in the provinces favored the abolition of government regulated prices.[48] The Finance Minister von Bodelschwingh, who also was the highest mining administrator in the 1830's, was even an opponent of state regulated prices. On the other hand, the lower echelon Prussian officials generally, and the leading official of the *Oberbergamt*, in particular von Oeynhausen, and his colleagues still favored the maintenance of the old tax price system. They argued, in true bureaucratic manner, that the tax price was "a means to maintain the necessary order and should not be undermined but rather used in an appropriate manner."[49]

C. Business Organization and Finance

1. The Mining Company and Joint-stock Company

The two major private business organizations in the Ruhr mining region, the mining company, _Gewerkschaft_, and the joint-stock company with limited liability, _Aktiengesellschaft_, answered different needs of the mining industry at different times. When mining was in its early stages of development and under the strict control of the _Direktionsprinzip_, the traditional and locally constituted mining company had been the suitable form for a private business organization with defined and restricted participation. As the mining industry developed, its needs changed. From the 1830's onward the crucial factor in the industry was the constantly increasing demand for capital.

The _Gewerkschaft_ was the major source and at times the sole source of money for the Ruhr coal industry. In the absence of financial institutions and a capital market, the mining companies acted as their own regional financial institutions, providing local capital for the development of the industry. The _Direktionsprinzip_ with its legal and administrative framework reinforced this function. It centered state control on the individual mining operation, thereby effectively reducing the company and its shareholders to the category of a "sleeping partner."[50]

The mining company under this system became the organization which provided the capital, while the state provided the management for the individual operation and the general administration for the entire industry. This partnership worked fairly well in the early stages of the industry's development.[51]

As the mining industry grew in complexity and size, demands increased on the mining company and its inadequacies appeared. During the first half of the nineteenth century, the <u>Gewerkschaft</u> itself became increasingly an obstacle in the growth of the Ruhr mining industry. It no longer was an effective vehicle for large scale mining; in part this was caused by the <u>Direktionsprinzip</u>'s restrictions on the organization.[52] Not only the legal restriction on decision making within the company, but the legal nature of the company and its shares were questioned.

There were several major criticisms of the mining company and its shares. The problems arising from the shares became particularly severe only after 1830.[53] These problems were generally known but there had been no agreement on the exact nature of the reforms. The major complaints voiced in the 1830's and after opposed: (1) the fragmentation of a share into unequal parts; (2) the slow and costly transfer of shares in a time of frequent changes in ownership of mining property; (3) the uncertain value of a share and the company as a whole; and

(4) the complexity of loans and claims on the share or its parts and the difficulties in accounting or the transfer and claims on the shares.

One of the major complaints was against the splintering of the share into uneven and unrepresentable parts, hindering sales and decision-making in the company. Already in 1792 the <u>Allgemeine Preussische Landrecht</u> had indicated a desire for reform. The relevant paragraph prohibited the division of a share into parts smaller than 1/8 but failed to emphasize the equality of the parts.[54] The result was that the share was divided into unmanageable fractions such as 3/17, 9/64, etc., as long as the fraction was more than 1/8. But in actual practice there were even much smaller fractions.[55] This made decision-making and voting in the company difficult, since only one full share had one vote.

Another major complaint was the immobile character of the share itself. C.J.B. Karsten, the greatest contemporary authority on mining law in the first half of the nineteenth century, argued for the "mobilization of the share." According to him the <u>Gewerkschaft</u> should be made similar to other business organizations:

> Only the mining operation or the <u>Gewerkschaft</u> itself should be listed in the public records as a single, indivisible and immobile entity, while the individual shareholder's part of the mining property should be regarded as part of his own mobile personal wealth and should be free to his unhindered disposal.[56]

Karsten argued that the taking on of loans or debts in

general by a mining company should be decided by all its shareholders, but that the private debts of the individual shareholder on his share should be entirely excluded from the listing of the company's assets and liabilities.

As it was, the very nature of the share lowered the credit rating of the companies, at a time when the large mines needed increasingly more capital. The basic reason for the lower credit was that the total value of the shares was quite different from the actual market value of the mine itself. For instance, a company whose financial position was unfavorable could possess in the mining property itself a credit security whose value could by far exceed the total value of its shares. Further, the mining operation could have debts or outstanding assets as a company but this would not be reflected in the shares. A share as an "ideal part of the company" did not reflect the present value of the company. On the other hand the individual shareholder could put up his share as a security for a loan, but this personal debt would also be listed on the company's books and depress the company's credit rating as a whole.

The security minded opponents of the "mobile share" feared it would lead to speculation. They argued that this speculation would have disastrous effects on the finances of the mining companies. They did not want to have the shares on "public auction"--on the open market. The proponent of the mobilized share argued that the reformed share would not be a public share. Unlike the shareholder in a joint-stock company with limited liability, the share owner in a *Gewerkschaft* could not buy with one single payment a permanent part in the enterprise. The shareholder was still continuously responsible for his part in the company; he not only shared in the profit but still had to pay *Zubusse* when required. Therefore, while the mobile nature of the *Kuxe* would benefit the capital needs of a company, the *Zubusse* would still deter speculating interests. Furthermore, as the defenders of the mobile share argued, some speculation was necessary; otherwise no one would take the risk of developing new mining fields.[57]

Despite the arguments the share system was not reformed until 1865. In the meantime this complicated and confused system made mortgaging of entire mining operations nearly impossible. Furthermore, the credit of the mining operation was considerably lowered by the personal debts of the shareholders on their shares.

The capital for the early shaft mines came from active entrepreneurs within the region. Despite the increasing demand for capital from mining, the financial institutions in the Ruhr region itself remained backward. During the first half of the nineteenth century a few private firms had been established, but all of these were small and dealt mostly in exchange and discount business.[58] Therefore the major source of capital for shaft mining came from the investment by merchants and coal dealers within the region. Such families as Stinnes and Haniel were the first to risk shaft mining. The Stinnes family had been actively engaged in the coal trade since the early nineteenth century. Mathias Stinnes, coal dealer and transporter, also participated in mining operations. When he died in 1846, he had shares in 36 mines. He had also been one of the first to take the risk of sinking a shaft through the water-bearing chalk layer at his mine, Graf Beust, in 1839. The Haniel family turned also to shaft mining in the 1840's and the mine Zollwerein finally started operation in 1851.[59]

Already in the 1830's but especially in the 1840's non-regional and foreign capital was attracted for the first time to Ruhr coal mining. The major reasons were the success of shaft mining, and in the 1840's the transformation of transportation due to the Ruhr region's connections with major railroads.[60] The law of November 9, 1843, also helped to attract capital. This law defined the legal foundation of the joint-stock company in Prussia and permitted them in mining and related industries. But with the direct permission of the king joint-stock companies had been allowed in mining since 1834.

The joint-stock company attracted non-regional and foreign capital because these mining interests were familiar with such an organization. The _Gewerkschaft_ was hardly known outside of the mining areas and not at all known in most of the foreign countries. Unlike the Gewerkschaft the joint-stock company in mining was not under the Direktionsprinzip. While the internal affairs of the organization were not under the mining law, the acquisition of mining property and the actual operation of a mine was under the mining administration's direction.

Before the end of 1853, there existed 27 joint-stock companies in all of Prussia engaged in mining or related industries. As Table 4.7 below shows, 11 of these owned or directly operated mines in the Ruhr region.[61] All of these had non-regional participation and all except one had capital of more than 500,000 thalers. Of course some of these, particularly after 1852, were ironworks which in addition owned coal mines. Furthermore, most of these joint-stock companies had more than 2,000 shares; the Gewerkschaft even at the legal limit of share fragmentation to 1/8 could only yield 1,024 parts. One of the complaints against the Gewerkschaft was that the individual share of a large mining operation demanded more from the typical investor than he was willing or able to invest.

The first joint-stock mining company in Prussia and in the Rhenish-Westphalian provinces was the Eschweiler Bergwerksverein, founded in August, 1834. Its seat was in Aachen and it was engaged in coal mining.[62] The first joint-stock company in the Ruhr region was the Hardenbergischen Kohlenbergwerke, established in March, 1840.

Its seat was in Düsseldorf; it operated mines but was in the process of liquidation in 1853.[63] The first joint-stock company with its seat in the Ruhr region in Essen was Concordia. The Haniel family had founded this company which was solely in engaged in shaft mining.

The first three joint-stock companies had their seats outside the Ruhr region in the Rhine cities of Cologne and Düsseldorf. Both of these cities were also heavily involved in the building of the first railways through the Ruhr. Entrepreneurs from Cologne became particularly interested during the 1840's when the Cologne-Minden Railway was being constructed. Here Mevissin and his friends, financiers and industrialists, bought in 1844 a number of mining claims and out of their consolidation in 1849 established the Kölner Bergwerksverein. The first shaft mine was started in 1847/48.

Although foreign capital was invested in the Ruhr, the first registered foreign joint-stock company was the Societe Anglo-Belge des Charbonnages du Rhin, (Englisch-Belgische Bergwerksverein) in 1848. This company was reformed to the Societe Anonyme du Charbonnage-Belge-Rhenan after the English capital pulled out during the revolutionary year of 1848.

TABLE 4.7

THE FIRST JOINT-STOCK COMPANIES IN THE RUHR REGION: 1840-1853

Name of Aktiengesellschaft	Seat	Year	Capital (Thaler)	Number of Shares	Value of Share	Purpose and Ownership
Hardenbergische Kohlenwerke	Düsseldorf	1840	640,000	2,400	266	Owned coal mines
English-Belgische	Düsseldorf	1848	666,666	2,500	666	Coal mines in Essen area
Kölner Bergwerksverein	Cologne	1849	200,000	1,000	200	Mines in Essen area
Concordia	Essen	1850	550,000	1,100	500	Mines in Essen area
Hörder	Hörde	1852	2000,000	10,000	200	Mines in Bochum area Ironwork at Hörde
Belgische-Rheinische	Düsseldorf	1852	533,333	2,000	267	Coal mines in Ruhr
Phönix	Aachen	1852	1500,000	7,500	200	Ironworks in Essen
Concordia Eschweiler	Eschweiler	1853	1000,000	5,000	200	Interests in Ruhr coal
Vereinigte Westphalia	Dortmund	1853	500,000	1,000	500	Mines and field claims in Ruhr
Friederich Wilhelmshütte	Mühlheim	1853	512,000	1,024	500	Coal mines and ironwork
Maassener	Dortmund	1853	1200,000	6,000	200	Owned fields in Dortmund

Source: "Akktiengesellschaften für Bergwerks-und Hüttenbetrieb in Preussischen Staate," Zeitschrift für Berg-Hütten-Salinen Wesen, Vol. II (1855), 102-109. See also note 64.

2. The Tax System Under the *Direktionsprinzip*

The greatest financial deterrent to the development of the Ruhr mining industry in the second quarter of the nineteenth century was the burden of taxation. The price-tax system effectively cancelled the earlier benefits conferred by the state. Originally a high and stable state income had been only one of the state's several goals. But during this period the major purpose of the state's control of prices, of entry and exit rights and of the management of mining operations appears to have been the collection of a tax revenue. Between 1830 and 1850, the sole purpose of the price system in particular seems to have been to regulate tax revenue, especially the *tenth*. In fact there was no economic rationale for demanding such high taxes from the mining industry since no other industry was burdened by them. But the state tended to use the *Gewerkschaft* as a convenient source for financing other governmental functions not benefitting the mining industry.[65] From the industry's point of view, the taxes became an increasingly serious obstacle to development, inhibiting the introduction of large scale shaft mining and frustrating the industry's increasing need for capital.

R. von Carnall, the editor of the journal *Zeitschrift für Berg-Hütten-und Salinenwesen*, published with the permission of the ministry and himself a high ranking mining official at Berlin, had already fought in the 1840's for the reform of the <u>Direktionsprinzip</u>.[66] Carnall was the most outspoken critic of the tax system, arguing vehemently for a tax reform, especially for a reduction of the <u>tenth</u>. He argued that the taxes were strangling the Ruhr mining industry.[67] More than anyone else, he influenced the reform of taxes as incorporated in the tax law of 1851.

Essentially, Von Carnall argued that the conditions in the 1840's had dramatically changed and that the price-tax policy of the <u>Direktionsprinzip</u> was outmoded. The railroads, the entrance of large shaft mines, and the increasing competition among mines had drastically altered the structure of the Ruhr industry, demanding a reform. For the 1848 revision he supplied statistical information based on the years 1843 through 1847, showing the distressing relationship between taxes and mining income.[68] During these years the Essen-Werden region, where nearly all the shaft mines were located, had the highest operating cost and also the highest taxes of any region in Prussia, consequently it had the lowest income.

According to von Carnall the average value of production in the Essen-Werden region for the five years 1843 through 1847 was 939,000 thalers, divided in the following manner: 733,000 thalers or 78.2 per cent for the cost of operation; 206,000 thalers or 21.8 per cent net income before taxes; 135,000 thalers or 14.3 per cent of the value of sales for taxes; income after taxes 71,000 thalers or only 7.1 per cent of the gross value of production and from this amount repayment of debts and payments of dividends had yet to be made. This meant that nearly two-thirds of the net income or 65.3 per cent went for taxes. Tables 4.8 and 4.9, which are presented at the end of this chapter, compare the operation costs, net income and taxes for Upper Silesia, Lower Silesia, Mark (Ruhr), and Essen-Werden (Ruhr) as well as those for Prussia as a whole.

They indicate that the cost of operation and the taxes were higher in the Essen than in Upper Silesia. Since the majority of mines in the Essen region were shaft mines, the taxes actually tended to discriminate against the new and large scale operations. The new method of production involved high risks, higher initial capital for the shaft and steam engines, higher operation costs, and a great deal of private initiative. Yet under the _Direktionsprinzip_ before 1851 this new and necessary form of mining was precisely discriminated against by taxing the profits away.

D. Mining Law and Its Administration

The orderly and efficient administration depended in large measure on the ability to administer the mining law. The law itself, however, made orderly application difficult and the administration was incapable of overseeing and administering the laws. The legal system itself was unclear and cumbersome.

In Prussia, the _Direktionsprinzip_'s legal expression was found in the form of twelve different mining codifications,

TABLE 4.8

TAX TABLE, BASED ON FIVE YEAR AVERAGE VALUE OF PRODUCTION 1843-47

Region	(1)	(2)	(3)	(4)	(5)	(6)	(7)
Upper Silesia	1,568	956	622	138	37	175	447
Lower Silesia	654	499	155	64	11	75	80
Mark (Ruhr)	1,425	1,096	329	125	68	193	126
Essen (Ruhr)	939	733	206	88	47	135	71
Prussia	7,157	5,140	2,017	490	224	714	1,303

(1) Five Year Average Value of Production (1,000 thalers)

(2) Cost of Operation (1,000 thalers)

(3) Net Income (1,000 thalers)

(4) The Tenth-Taxes (1,000 thalers)

(5) Other Taxes (1,000 thalers)

(6) Total Taxes (1,000 thalers)

(7) Income after Taxes (payment of dividends or repayment of debt still included)

Source: Carnall, Die Bergwerke in Preussen und deren Besteurung, pp. 65, 66.

TABLE 4.9

TAX TABLE, PERCENTAGES OF TABLE I

Regions	(1)	(2)	(3)	(4)	(5)
Upper Silesia	60.5	39.5	11.1	28.4	28.1
Lower Silesia	76.3	23.7	11.5	12.2	48.0
Mark (Ruhr)	77.0	23.0	13.1	9.9	58.6
Essen (Ruhr)	78.2	21.8	14.3	7.1	65.5
Prussia	72.0	28.0	10.0	18.2	35.4

(1) Cost of operation from value of production

(2) Net income from value of production

(3) Taxes from value of production

(4) Income after taxes from value of production

(5) Proportion of taxes to net income

Source: Derived from H. von Carnall, "Besteurungen..." 1850, pp. 65-66.

and Part II of the Allegemeines Preussisches Landrecht. In the Ruhr mining region alone, there were three different mining codifications after 1815 plus the Landrecht, as well as a host of special privileges dating back to feudal times. The result was a splintering and confusion of legal provisions concerning mining operations. The legal confusion was increased not only after 1803 when Essen-Werden fell to Prussia but particularly after 1815 other small territories were added each with their own mining tradition.

The mine owners on the princely territory of Broich fought a long legal battle to stay out of the Prussian system.[69] This territory, including Mühlheim and the surrounding area, had its own Bergregal. After 1815, it fell to Prussia. The mining companies had great freedom, especially at the management level. Consequently, when on May 5, 1816, the Direktionsprinzip was to be installed, they fought against it. One of the mining companies, Wiesche, even publicly claimed that it could operate better under the management of Franz Dinnendahl than under the mining officials. Dinnendahl, after all, was one of the great leaders of the Ruhr region.

Despite legal pressure, the mines in the Muhlheim area successfully fought the introduction of the Direktionsprinzip for nearly 30 years. Even a special cabinet order of 1834, ordering the companies to be brought under the system, appeared to be ineffective because as late as 1843 they were still protesting when the mining office tried to enforce it. Finally, in 1845, the two remaining independent mines, Wiesche and Sellerbeck, were forced to accept state management.

Ironically, the Ruhr mining administration introduced the *Direktionsprinzip* just 6 years before state management of individual firms was abolished. Both mines, Wiesche and Sellerbeck, were among the largest in the Ruhr region. In 1850 Sellerbeck was the fifth largest and Wiesche the thirteenth largest mine in the Ruhr. Certainly both mines had illustrated that private enterprise could successfully manage its own firms without the help of the mining administration.[70]

The major cause for the chaos and the inability to administer the mining law by the 1830's, however, was the complicated procedures required to gain entrance into the industry. In the last quarter of the eighteenth century many *Mutungen* were granted to speculative interests, who foresaw the future growth of the mining industry. During this period, however, there were also excellently trained mining officials, and capable leaders who understood how to direct the mining industry. These officials tended to be practical and liberal in the interpretation of the laws. For the most part these liberal practices were continued until 1817, the end of the term for the *Oberbergmeister* Bölling, head of the *Oberbergamt* at Dortmund. But during the following period until the early 1830's, the number of warrants for prospecting, claims and other application accumulated without action.

There was much discussion among the mining officials, but no agreement. This was mainly because the existing laws were unclear and arbitrary interpretations were common. In the years 1830-34, moreover, mining enjoyed a great boom as a result of free trade with Holland.[71] The old applications still had not been acted on, and new ones were coming in by the hundreds. In addition, profitable speculation was going on with old claims.

In an 1833 report, Honigman, a mining official at the mining office at Bochum described the prevailing conditions in the Mark mining region which was under the jurisdiction of the Bochum mining office. "Some reform is absolutely necessary." Describing the state of affairs in 1832, he explained, "there were the following applications to be acted on: 287 prospecting permits, 206 claims, 116 claims awaiting inspection, and about 1500 already existing applications of all types. Many were even dated prior to 1800."[72]

From more than 2,000 applications, action was taken on only four. Three field concessions were granted and the official measuring of the field for the fourth application was taken. But the net result was only one actual operation. Gottessegen and Erbstollen, an old mine, was reopened. For sixty years its owners had paid _Rezess_ taxes to keep possession of the field.

Honigman complained that the law of July 1, 1821, merely added to the general state of confusion in the Mark mining region. "It was supposed to be applied and every one strives for this today."[73] The law increased the size and depth of the mining fields and was intended to help shaft mining, allowing larger fields to operate more economically. As already mentioned in Chapter II, the law was badly formulated, and its uncertain technical aspects hindered its application.

According to Honigmann in an area of 2 1/2 square miles in the Mark region, "there already exist 91 tunnels used for drainage with rights to mine the coal in them, 917 single field concessions from which only 137 are mines in operation, and finally there are 1500-2000 claims." He asked, "How could it, therefore, be possible to allow further prospecting and claiming of fields according to the larger fields of the 1821 law?"[74] He thought this law was totally inapplicable. Instead of clarifying the existing confusion, it created even more. Instead of helping, it hindered.

The cabinet order of June 4, 1835, was a stop-gap measure to avoid even further confusion.[75] It prohibited new claims in two major areas in the Ruhr mining region. It admitted the existing confusion by stating "in those areas in the Mark <u>Bergamt</u> district where the number of field concessions and claims makes it difficult to recognize the remaining free fields, no new claims are allowed. First all old claims have to be settled."[76] It provided a breathing spell for the overloaded mining offices and mining courts which were trying to deal with the applications and settle the many legal disputes. This suspension was not lifted until the cabinet order of January 21, 1853.[77]

FOOTNOTES - CHAPTER IV

1. Henderson, (on Vincke), pp. 56-57; additional information, pp. 48-57.

2. A.M. Prym, Staatswirtschaft und Privatunternehmung in der Geschichte des Ruhr bergbaus, (Essen 1950), pp. 84, 85.

3. R. Klostermann, Lehrbuch des Preussischen Bergrechts, p. 41. Gives the cabinet order concerning the Schölerpad company.

4. H. Spethmann, "Der Tiefbau der Zeche Schölerpad im Kampf gegen das Direktionsprinzip," Bergfreiheit (1952), Section 2.

5. Wilhelm, Aus der Geschichte des Steinkohlenbergbaues in Bommern a./d. Ruhr. (Mining Library Essen Nr. Fa 57) Photocopy.

6. Spethmann, Die geschichtliehe Entwicklung des Ruhrbergbaus um Witten und Langendteer, (Gelsenkirehen, 1937), pp. 55-56.

7. Spethmann, Die ersten Mergelzechen im Ruhrgebiet, (Essen 1947), hektographiert, (Mining Library, Essen), pp. 25-27.

8. Ibid., pp. 108-109.

9. Die Geschichtliche Entwicklung der Gewerkschaft Zeche Heinrich in Überruhr, (Essen, 1834) Mining Libr. Essen.

10. Bericht vom 22. Febr. 1843, (O.B.A.-84/17) Münster Archive.

11. Reuss, pp. 63-65.

12. Reuss, p. 65.

13. Spethmann, Ruhrgebiet, p. 159.

14. Ibid., pp. 160-161.

15. Willy, Berölkerungsbewegungen, p. 26.

16. Spethmann, p. 150.

17. Henderson, p. 3 n.

18. Reuss, p. 14.

19. *Z.f.B.H.S.-W.*, vol I, pp. 36-38.

20. Willy, pp. 22-26.

21. Däbritz, "Entstehung und Auf bau des rheinisch-westfälischen Industriebezirks," *Beiträge zur Geschichte der Technik und Industrie*, (Berlin, 1925), p. 106.

22. Reuss, pp. 55, 56.

23. Däbritz, p. 36.

24. Däbritz, pp. 38, 39.

25. Carnall, *Die Bergwerke in Preussen*, p. 56.

26. Krampe, p. 106.

27. *Ibid*., p. 107 n.

28. Carnall, *Besteurung*..., p. 50.

29. Ludwig Müller, *Kohlenverkauf auf den Ruhrzechen im Essen schen*, part I (Bonn, 1838); Ibid, part II (Bonn, 1840), p. 10, mining Libr. Essen.

30. Krampe, *Staatseinfluss*, p. 89.

31. Krampe, p. 103.

32. Krampe, p. 103.

33. Müller, pp. 58-59.

34. See 29.

35. Müller, p. 25.

36. Müller, p. 14.

37. Müller, p. 17.

38. Müller, p. 17.

39. Müller, p. 19.

40. Müller, p. 20.

41. Müller, p. 25.

42. Müller, p. 25.

43. Müller, p. 25.

44. Krampe, Staatseinfluss, p. 90; based Statuten der Essen-Werdenschen Stein-Kohlen-Handlung (1838), O.B.A.-C/43, Münster Archive.

45. Der Deputierte von Vereinigte Kapellenbank und Clarbeck an das Oberbergamt, 18. Sept. 1838, (O.B.A.-165) Münster Archive.

46. Krampe, pp. 99-100.

47. Zeche Langenbrahm in ihrer Entstehung und Entwicklung 1772-1922, (Essen, 1922), pp. 104-107.

48. Carnall, Bergwerke in Preussen..., pp. 50, 51.

49. v. Oeynhausen an v. Milicki, 18. Der. 1843 (OBA-C/42) Münster Archive.

50. Boldt, Staatsbergbau, p. 3.

51. Otto Veit, Der Bergbau als staatliche Finanzquelle, (Diss. Frankfurt, 1927).

52. W. Däbritz, "Finanzprobleme aus der Entstehungs zeit des rheinisch-westfälischen Industriereviers," Glückauf, (1922), pp. 1355-1376.

53. W. Däbritz, "Entstehung und Aufbau...," Beiträge, III (1911).

54. Klostermann, "Bemerkungen über die Mobilisierung der Kuxe," Z.f.B.H.S.-W., (1861), p. 315.

55. Däbritz, p. 22.

56. Klostermann, p. 316.

57. The argument contained in the above two paragraphs is based upon Klostermann, 316-323.

58. Spethmann, Ruhrgebiet, vol I, p.

59. Däbritz, p. 33.

60. For French capital in German coal see Rondo Cameron, _France_, pp. 371-375.

61. "Aktienvereine für Bergwerks-und Hüttenbetriebe im Preussischen Staate," _Z.f.B.H.S.-W._, II, 102-109.

62. "Eschweiler Bergwerksverein," _Z.f.B.H.S.-W._, II (1855) p. 102; Cameron, p. 372. This A.G. had French capital participating.

63. "Hardenbergische Kohlenbergwerke," _Z.f.B.H.S.-W._, II, 103; Cameron, p. 373.

64. Nearly all of these A.G.s had foreign capital participating. See Cameron, pp. 372-375, for the more important A.G., having French capital participating or being solely owned by French businessmen.

65. Otto Veit, _Der Bergbau als staatliche Finanz: quelle_, (Diss. Frankfurt, 1927).

66. See R.v. Carnall, _Übersicht von den bisherigen Arbeiten der zur Entwerfung eines neuen Berggesetzes ver sammelten Commission_," Sept. 1848, Berlin.

67. Carnall, _Die Bergwerke in Preussen und deren Besteuerung_, (Berlin, 1850).

68. Carnall, _Besteu erung_, pp. 65-66.

69. Oberbergamt Dortmund, "Widersetzlichkeit verschiedener Gewerke gegen die im Jahre 1803 im Essen-Werdenschen geschehenen Einrichtungen der Verwaltung des Bergwerksregal, 1803-1806," O.B.A. Dortmund, B.Gr. 3, Nr. 11.

Reuss, "Mitteilungen...," p. 61.
Other mines fought long legal battles; nearly all involved the refusal to pay taxes, esp. the Tenth, O.B.A. Dortmund, B.Gr. 55, n,r,l,m, (1828-1837).

70. "Bergwerke in Preussen," _Z.f.B.H.S.-W._, Vol. I, pp.

71. Oberbergamt Dortmund, "Steinkohlenhandel nach Holland ...1812-1838," O.B.A. Dortmund, A Nr. 466; Reuss, "Mitteilungen...", p. 62.

72. K. Oberste-Brink, "Die Vierung...", _Glückauf_ (1941) pp. 659-53, 667-684. In this article hereprints the entire report of Jan. 26, 1833 by Bergmeister Honigmann." Das angeordnete und verwickelte Bergrechtsanwesen auf den Steinkohlen Flötzen der Grafschaft Mark," pp. 651-53.

73. *Ibid.*, p. 652.

74. *Ibid.*, p. 652.

75. *Ibid.*, pp. 652, 53.

76. Oberste-Brink, p. 653.

77. *Ibid.*, p. 653.

CHAPTER V

LEGAL REFORMS OF THE *DIREKTIONSPRINZIP*

In the eighteenth and for much of the nineteenth century conflict between various interest groups frequently led to a stalemate in state policy-making, to ineffectual implementation of policies, or to very slow progress. In addition, foreign influence, significantly that of France in the early nineteenth century, affected Prussian policy.

The first section of this chapter considers the various interest groups and their relation to proposed reforms and discussed the effect of the Napoleonic era on the Ruhr region. The second section deals with mining law reform, specifically the unsuccessful revision attempts from 1826 to 1851. The third section analyzes the reform laws between 1851 and 1865. The last section deals with the opposition to these reform laws by the mining offices, particularly the *Oberbergamt* Dortmund.

A. General Prussian Reforms

Prussian state policy in the eighteenth and much of the nineteenth centuries was not necessarily consistent, uniform, or effectively applied. The various socio-economic and political interests in the state were frequently reflected in the making and application of state policy.[1]

The existence of certain major social, political, and economic groups with conflicting interests crucially influenced Prussian policy and its effective implementation. Among these groups were the Junker class of the eastern Prussian provinces, the king and the leading civil servants, the tradition-oriented state bureaucracy, the merchant class of the older cities, and the new industrial interests gradually gaining importance, particularly in the western provinces.

These groups were by no means clearly defined. Both the groups and individuals from these groups were interrelated if not through social class or geography than through common economic or political interests. For instance, the Junkers were members of the dominant group in Prussia, the aristocracy, and as large estate owners represented the strongest political and economic interest in eastern Prussia. The aristocracy of the western part, especially of the Rhenish-Westphalian region, had to share its power with the bourgeoisie. In fact the dominant interest group in the Rhineland was the merchant class of the old cities.

The Junker class provided the officers for the Prussian army as well as the bureaucracy and leading civil servants. These aristocrats as members of the state bureaucracy, gradually shifted their primary loyalty from their own class to the state. Certainly by the end of the eighteenth century the leading state officials had become devoted civil servants of the Prussian state and were the king's closest advisors. For instance, in the mining administration aristocrats served as leading administrators and bureaucrats. Furthermore, the mining administration's bureaucracy developed its own loyalty, tradition, and philosophy. This was to cause conflict in the reform of the Direktionsprinzip.

Prior to about 1850, Prussia was primarily an agrarian state; consequently the landed interests were among the most powerful. Therefore, the role of the Junker is central to any analysis of Prussian state policy. These large-estate owners of the eastern provinces, were the backbone of the state. Even Frederick the Great, an exceptionally progressive minded and powerful king, had to bow to the Junker class. A proverbial piece of Junker philosophy was, "the king is all powerful and may do as he pleases, as long as he does not interfere with our business." For example, Frederick the Great in the 1770's attempted some agricultural reforms which contained the seeds of the great reforms of the early nineteenth century. According to Sombart, he ran into such strong opposition from the Junkers that he was only able to carry them out to a limited degree on his own estates.[2]

The Junkers as a class frequently represented a negative force in the early economic development of Prussia. They were in principle against any change affecting their position in the overall social and economic structure of the Prussian state. Consequently they were against early forces of industrialization and eyed suspiciously any reforms that would strengthen the growing industrial interests. Further, there was a tension between the predominantly agricultural eastern provinces and the more industrialized western parts of Prussia. The agricultural east watched with great concern the gradual territorial expansion in the west bringing older and more industrialized regions with different social and political structure into the Prussian state.

By the beginning of the nineteenth century further tension was created when two economic philosophies appeared to clash. The liberal economic ideas of the Physiocrats and the English classical economists, principally those of Adam Smith, came in conflict with the economic and political ideas of mercantilism and Prussian enlightened despotism.

Mercantilism had imposed personal constraints, collective forms, and collective regulation and regimentation. According to the new philosophy, the individual should be free from personal constraints, free from collective regimentation. The state should not interfere with the market. Competition would be the regulating force. Individual responsibility and private initiative in all economic activities should be fostered.

The liberal ideas did have an effect on Prussian policy; they influenced the content of the general reforms after 1807. But the immediate cause for reform was political.

Prussia's defeat at Jena by Napoleon and the humiliating peace of Tilsit in 1807 was the major reason for the state's willingness to reform. Economic and social conditions were desperate. The government was bankrupt. It needed public support to fight Napoleon and was willing to bargain for it. Above all an attempt was made to draw the educated and moneyed elite into active participation in the life of the state.

In October 1807 Freiherr vom Stein became chief government minister and immediately pursued a policy directed to gain the support of the general citizenry. Stein had concluded that the defeat at Jena was due to the failure of the Prussian monarchy to avail itself of the general public support by not having treated the population as a partner. Vom Stein abolished serfdom and many feudal laws (though retaining manorial jurisdiction and some peasant feudal obligations). He established the civil equality of all citizens and reformed local government to permit ordinary citizens to participate in it. Finally he reformed the military by creating a reserve system which provided the army with trained manpower in case of war. After vom Stein was forced out of office by Napoleon, Chancellor Hardenberg continued reforms between 1810 and 1817. He was particularly responsible for streamlining the civil service by centralizing it.

Although caused by military and political necessity,

the reforms used the new economic philosophy as a convenient vehicle. The introduction to the famous Emancipation Edict of October 9, 1807, was couched in liberal political and economic terms. The king admitted the disastrous consequences of the defeat by Napoleon and acknowledged that the state had become incapable of continuing its previous economic and social policy of regulation and leadership. It lacked the finances. For the sake of "general welfare and justice" he promised "the removal of everything which until now has impeded the individual in the pursuit of that level of personal well-being which he is capable of reaching according to his own capacity."[3]

At the same time, the king also promised greater freedom to private enterprise in the mining sector. The possibility of changes in the patriarchical state control of this industry, at least a liberalization of the Direktionsprinzip, appeared to exist. One of these favorable signs came also after Prussia's defeat by Napoleon. The king himself in the cabinet order of April 15, 1809, appeared to foreshadow a new attitude:

> Henceforth, I do not wish to see private industry in the operation of privately owned mines as restricted as it has been heretofore. The state should not engage in new mining projects on its own account, except in those special cases where it would benefit all concerned.[4]

In addition, an instruction from the king of July 27, 1809,[5] declared that mining should be open to all interested persons. Accordingly the mining administration should not withhold prospecting permits, granting of claims, and field concession even if it conflicted with older law. Article 8 of the same instruction, however, stated that mining companies should respect the regulation of the mining administration of their technical and business operation. But these were to be the only limitation on private enterprise. It was not followed up by concrete instructions. Of course, at this point Prussia had lost its western territories, including the Ruhr mining region, to France and this directive was focused on the mining activity on Silesia.

Although the reforms promised great changes, including the state's attitude toward the social and economic life of the country, in most cases the changes failed to materialize immediately. Shortly after the Congress of Vienna a general period of conservatism followed. In Prussia the king dismissed the Diet in 1820 and it did not recall it until 1847. In general the reforms failed to be applied effectively and were continued only in the 1850's and 1860's after a break of thirty to forty years. The agrarian reforms failed because of the opposition of the Junker class.[6] Other reforms involving the basic freedoms of occupation, contract, and property were successfully continued after 1850. This was certainly the case with the mining law reform. Although revisions were attempted earlier, the Direktionsprinzip was not reformed until the 1850's.

In the case of the western territories and particularly the Ruhr region, the changes that did take place in the beginning of the nineteenth century were not due to the Prussian state but rather were the result of French rule. After the defeat of Napoleon Prussia regained its western territories, but found that French rule had brought about some fundamental changes.[7] A brief assessment of French rule from 1807 to 1815 indicates both positive and negative effects on the political and economic welfare of these territories. An immediate effect of the French rule was the political unification of the numerous splintered political entities. Before the French, there had existed in the later Prussian province of Rhineland fifteen ecclesiastical and seventy-five secular rulers, as well as the two imperial cities of Cologne and Aachen. In the case of the future Province of Westphalia, the counties of Mark and Jülich, the Duchies of Münster and Berg, and the territories of Essen and Werden were unified under French rule.

French rule had effectively removed a major obstacle to the development of the western territories, political fragmentation. When Prussia regained these territories it could easily reconstitute the three French departments into the two provinces of Rhineland and Westphalia. The French also removed many of the traditional feudal obstacles to economic growth. The church tithe, special aristocratic privileges, feudal burdens, and guild restrictions were abolished.

With political unification under the French came also economic unification. On December 8, 1807, the French government ordered the abolition of all internal tariffs and established a unified external tariff system for the entire territories. Napoleon should also be credited with the improvement and construction of roads. For instance, in the Duchy of Münster there had not been a single paved road; Napoleon built one.

Although French rule had many beneficial consequences, Napoleon's political involvement in Europe also adversely affected the Ruhr region. Napoleon recruited soldiers for his wars, quartered troops, collected high taxes and especially disturbed the social and economic life of the cities. The Continental blockade also adversely affected the Ruhr mining region. It was cut off from Holland and the sea by a high tariff barrier. Until 1811 the French government prohibited the export of coal to Holland. Only French coal could be sold in the Netherlands, which did not include coal from the territories on the Right Bank of the Rhine. However, at the vigorous protest of the Ruhr region the prohibition was lifted on December 12, 1812.[8]

The Ruhr mines came under the French Department of Mines, which did not change much of the Direktionsprinzip. Mining taxes remained high. French mining officials, however, did encourage the introduction of new technology, particularly steam engines. The first shaft mine, the Vollmund, with a 116 meter shaft, used a steam engine. In the same year, 1808, Franz Dinnendahl, known as the Mechanikus, placed his self-constructed steam engine in the mines Sälzer-Neuak. This was the first machine which was also used for lifting of the coal rather than just pumping water. More steam engines were introduced between 1808 and 1815, marking the beginning of shaft mining in the Ruhr region. However, these mines did not penetrate the water-bearing secondary rock until the late 1830's.[9]

During the French rule the Code des Mines[10] was introduced on the Left Bank of the Rhine. This codification was to guide Prussian reformers who were attempting to establish a unified Prussian mining codification. On the other hand, when the actual Prussian reforms of the 1850's did come about, they went much farther in fundamentally changing the state's relationship to private enterprise in mining.

Just as the Prussian king had used the liberal economic philosophy as a tool of propaganda, so did Napoleon and his advisors. Napoleon himself, during the hearing on the new code, explained:

> Since mining is free in England, why should it not be so in France? It is a grave mistake, when a government wants to be too patriarchical. Through its own zeal, it destroys freedom and prosperity.[11]

Count Girardin, presiding over most of the hearings, expressed an even more liberal philosophy: "The basis for a good and durable culture is founded on the strength of private interest." He then continued as a disciple of economic liberalism:

> ... freedom given to proprietors and owners of land results in much good and little evil; the public interest is best safeguarded when it has instead of an enemy, a fellow citizen privately interested and motivated.[12]

Under the French code the state was at least in theory limited to the exercise of its Hoheitsrecht, the right to grant field concessions because the state owned the minerals. Further, supervisory functions by the state included the supply of technical personnel to the mining companies in return for a special mining tax. However, as Girardin pointed out, the role of the technical personnel should ideally be that of advice and counsel to the mining operators and should not be construed as orders.

In fact, the French code in application was legally similar to the Direktionsprinzip with one important exception: the mine owners could function as autonomous entrepreneurs. Furthermore, the Code des Mines gave France a unified mining law. Whether these provisions were beneficial to French mining at this time is another matter, but this crucial weakening of the Direktionsprinzip was the goal of German and Prussian reformers. Compared to the Prussian mining law at this time, the French code was considered highly progressive. It became the ideal for mining reform in the Ruhr, especially since it remained in force on the Left Bank of the Rhine after 1815, serving as a constant reminder of the benefits of a unified mining law.[13]

B. Unsuccessful Attempts at Mining Law Reform, 1826-1851

Between 1826 and 1851 several general mining law revisions were attempted.[14] These revisions had the following objectives: (1) a general unified Prussian mining codification, (2) abolition of the *Direktionsprinzip*, (3) reduction or abolition of the special mining taxes and levies, (4) abolition of the right of the state arbitrarily to reserve any field for its own mining.

Seven reform proposals were drawn up, but all failed eventually. The need for a unified mining codification was admitted by all concerned, but there were sharp differences as to the future role of the state. Therefore, all the proposals can be classed as pro-*Direktionsprinzip* or anti-*Direktionsprinzip*, i.e. reduction of the state's role to supervisory and police functions.

The turning point for reforms came with the cabinet order of July 24, 1826. The Ministry of the Interior (*Ministerium des Innern*) was delegated to work out an improvement and revision of the mining law. But a total and basic change in the law was not to be considered. In April, 1829, the first unprinted revision was handed over to the Justice Department. It did not basically deviate from the *Direktionsprinzip*, but at least it unified the many codifications. Taxes and property were to be treated in a separate regulation which the Justice Department was to work on. It failed to do so, and the whole revision was eventually dropped. But a commission, comprising members from both departments, was established to continue the work.

The commission did provide the first printed revision of a general Prussian mining codification in August, 1833. It had separate instructions for the enforcement of the *Bergregal* as well as special additional regulations for the various provinces. It was revolutionary in both content and scope. It would have totally rejected the *Direktionsprinzip* and restricted the role of the mining offices to a general inspection and supervisory function. The offices were to regulate mining activities and properties and to protect the public safety in accordance with the technical and engineering knowledge of mining. The management of the mining operations, employment, and wages were left to the mining companies. The basic idea of this proposal was eventually embodied in the 1865 General Prussian Mining Codification. The proposal itself was sent to the various departments and again revised.

In 1835, the second printed revision was made public. The Justice Department wanted it sent first to all the *Oberbergämter* for further consideration. Basically this proposal was similar to the first. But during 1836-38 the mining headquarters could not agree. They objected to the reduced status of their offices.[15] As a result, the proposal underwent further revision.

The third official version of a new codification came out in 1841. It was a complete reversal, re-establishing the _Direktionsprinzip_. In the same year, this revision was presented to a meeting of the _Stände_ from Silesia, Saxony, Westphalia and the Rhine province. They endorsed the _Direktionsprinzip_ and function of the mining offices, but all favored greater freedom for the private sector.

The cabinet order of February 28, 1842, put von Savigny, the Minister of Justice, also in charge of a newly created special sub-section for legal revisions. In May, 1845, he presided over a commission which produced the fourth printed codification proposal of 1846. Again the role and the power of the state were preserved under the _Direktionsprinzip_. This revision was still under consideration when the revolution of 1848 occurred.

The political upheaval of 1848 changed the nature of the attempts at mining law reform. For the first time, mining and industrial leaders, the vitally interested parties, were allowed a voice in the new and fifth official revision of December 1848. During this year a large number of petitions had caused Minister of Commerce von Milde to call in a new commission of mining officials and entrepreneurs, including Franz Haniel and Friederich Harkort.

Minister Milde asked this commission to produce a law that would be timely but that would also consider the future development of the industry. The final result was a codification similar to the one of 1833, abolishing the Direktionsprinzip.[16] For the first time also there was a deliberate attempt to integrate the French Code des Mines with the traditional German mining law. The French law influenced the codification's interpretation of legal concepts and also its emphasis on private property and individual rights.

This revision was investigated by a commission composed of ministers and representatives of various agencies. It was slightly modified, superficially reintroducing the Direktionsprinzip, especially at the level of the provincial mining headquarters, but essentially remained unchanged.

In 1850, it was sent to the Prussian legislature for official passage. This revision did not get on the calendar of the law makers.[17] The following reasons were given. There had been too much work during the legislative year 1850/51. The 1850 revision was not favorably received by all mining operators. There appeared to be enough dissenters to warrant further investigation into all aspects of the codification. It also would be beneficial to watch the reaction to reform in other countries. Austria was revising its mining law. In fact, the government was still not ready to concede the necessity for an overall reform.

In 1865 a report to the lower house gave an explanation for the failure to provide a unified Prussian mining codification in the last 40 years. A major reason was the conflict between the high civil servants in the mining bureaucracy and the mining-industrial interests:

> The higher echelon of the bureaucracy wanted to uphold the old mining laws at all cost and it possessed the power to do so. All reform proposals had to come to them. In addition, these officials did not realize that conditions throughout Prussia varied. The less economically advanced regions were, of course, opposed to any change.[18]

For obvious reasons, the report failed to add that the ultimate responsibility for the failure rested with the state itself.

C. Reform of the _Direktionsprinzip_: Reform Laws, 1851-1865

Despite the increasing divergence between the objectives of the _Direktionsprinzip_ and the actual economic and social conditions in the first half of the nineteenth century, the reform movement's goal, a general Prussian mining codification, failed to be achieved. But both the revision attempts and the political tension of the late 1840's finally forced reform of the _Direktionsprinzip_. The reform begun in 1851 did not result in a total legal reform of the _Direktionsprinzip_ but rather through a series of laws abolished the basic principles of the old system. Certainly the state's attitude toward the mining sector was changing. A few basic laws released the mining industry from the burdensome state direction and regulation enforced by the mining administration.

Between 1851 and 1861, three major laws fundamentally

reformed the _Direktionsprinzip_ and reduced state control over the mining industry to a minimum. These laws essentially changed the relationship between the private mining company and the state's mining administration. These were: (1) the law of May 12, 1851, shareholder law;[19] (2) the tax law of May 12, 1851;[20] (3) the law of May 21, 1860,[21] granting freedom of contract to both mine worker and employer.

The first law dealt with the legal relationship among the shareholders of the private mining company, _Gewerkschaft_, and with the relationship between the _Gewerkschaft_ and the state mining administration. The second law simplified and reduced taxes, but also changed the nature and character of the special mining taxes. The third law concerned itself not only with the safety and police function of the mining administration but also dealt with the relationship between the mining company and its employees. It established contractual freedom between employer and employee. These laws generated all other laws in the period 1851-1865. Furthermore, they not only reformed the _Direktionsprinzip_ but also became the foundation for the General Prussian Mining Codification of June 24, 1865.[22] These laws and their intentions made this codification somewhat belated and anticlimatic.

The two laws of May 12, 1851, the shareholder law and the tax law reformed the most burdensome controls of the _Direktionsprinzip_. They effectively provided an immediate and direct solution to major demands of the reform movement in the previous twenty years: (1) private management and control for both the mining company and the mining operation; (2) simplification and reduction of taxes.

The tax law reduced the long fought over tenth to the "twentieth" as of the third quarter of 1851. Now a mining company paid only 5 per cent of the value of gross production at the mine. In all 24 special mining taxes were abolished. Seventeen of these special taxes applied to coal mining and were collected in the Ruhr region. The tax for possessing a non-operational mine was reduced to one thaler per year. Instead of the other 24 taxes a "supervision tax" was introduced and collected from all mining operations. It amounted to one per cent of the value of sales by a mine. In the same instruction, iron ore (Blackbank) found among the coal layers in the Ruhr region was tax exempted for six years.[23]

The taxes, the initial tenth reduced to 5 per cent in 1851, were still further reduced by one per cent each year between 1862 and 1865.[24] Effective January, 1865, this tax was only one per cent. Other special mining taxes paid to the mining administration for certain kinds of services to the mining companies were entirely abolished. The tax system and its economic effects will be discussed in the next chapter.

The shareholder law was by far the most important reform. It was the cornerstone of the reforms between

1851 and 1865. This law fundamentally changed the basic principles of the Direktionsprinzip and clearly indicated the intentions of the reform. It made the private mining company a viable economic organization. It gave the decision-making power over all important matters to the company and instead of the state controls under the mining administration, allowed the nearly unrestricted market competition to be its regulator.

Although, as we have seen in the previous chapter, the complex controls of the Direktionsprinzip had been breaking down, the laws had remained unchanged. Furthermore, even if since the 1830's these laws were circumvented by private enterprise or sometimes only reluctantly enforced by the mining administration, some essential laws concerning the mining company and its operation, as well as the entire industry, had to be obeyed. Only a change in the laws could effectively free the forces of private enterprise.

The shareholder law transferred major controls of production and sales from the mining administration to the private mining company and its shareholders. However, some controls were maintained by the mining administration. First we will discuss those controls which were transferred to the mining company. Not only the law itself, but its many instructions, ministerial orders, and regulations effected this transfer of authority.

Decision making within the company was made easier.[25]

Prior to this law important decisions concerning the mining property required a unanimous vote. Now a simple majority of the shares was sufficient for most normal decisions. But a three-fourths majority of shares was necessary when leasing the mining operation or consolidating it with another. Previously, unanimity was required. This new provision considerably facilitated consolidation of fields and operations during the 1850's. However, a unanimous vote was required for the sale or disbandment of the company.[26] In all cases the individual shareholder had recourse to the courts for an appeal.

The 1851 law required the shareholders to elect an official representative or a board. Each mining operation which was owned by more than one person had to elect a representative to a board of two to five members. They legally represented the company in all dealings with other companies and with the mining administration.[27] But most important these company officials had full responsibility for the management of the mining operation itself and were held accountable for it by the shareholders and the mining administration.[28] Prior to this law state officials employed by the company had been responsible for the actual mining operation. Now the privately elected representatives were fully responsible for the employment of qualified technical and business personnel.

Instead of mining officials, the representatives of the

private mining companies decided the total number of workers employed at their mining operations. They could directly employ non-union workers and determine their wages, but had no power over the members in the miner's guild, *Knappschaftgenossen*, and their wages. The private representatives also could work out their own technical plan of operation according to which the actual mining operation had to take place. The representatives of the companies were also responsible for the companies' internal finances, such as the collection of *Zubusse*, and paying out profits. Finally the representatives could also hold the same position for several companies with the shareholders' approval.

After the 1851 law most of the controls which the mining administration maintained were focused on the mining operation itself. Particularly the technical aspect of the mining operation remained under the close supervision of the mining administration.[29] Both the approval and enforcement of the technical plan of operation and the employment of qualified management personnel by the private companies demanded the exercise of the police and inspection functions retained by the mining offices.

An instruction accompanying the shareholder law emphasized the right and the duty of the *Oberbergamt* and its mining offices at all times to inspect the individual mining operation. If such inspection revealed inadequacies in the technical operation of the mine their correction was immediately required.[30]

The plan of operation had to contain all necessary information concerning a proposed operation.[31] This plan had to be approved by the mining offices. It had to be checked in accordance with the standards of existing mining knowledge for the safety of the operation in regard to the surface, and the life and health of the mine workers. For this reason the mining office maintained the right to inspect the mining operation and police the application of this plan.

This technical plan was required before an operation could start. If the private company did not or could not prepare such a plan, the mining office was required to do so. Once a plan was approved, it had to be followed, or the mining office had the right to close the operation. In case of emergencies or mining disasters the mining office had the immediate police power to close the mines and direct the rescue operation. The costs of such operations were paid by the company.[32]

Although the private mining companies could employ their own managerial personnel, the mining administration had the right and duty to check their qualifications. All technical, business, and bookkeeping personnel had to prove their qualifications through an official examination. If the mining administration rejected them as unqualified, it would appoint a qualified person.[33] During the transition period before all mining operations could find qualified personnel to manage their operation, the *Oberbergamt* was allowed to supply the necessary personnel from its own ranks.[34]

Although the private mining company was free to determine the size of its labor force, the mining administration controlled the employment of guild miners and their wages. The representative of a company together with the mining office determined the number of workers needed for the operation. The company was required to employ guild miners before non-guild labor. This was particularly enforced during times of unemployment. Finally, in any dispute over the wages of non-guild labor, the mining office had the right to intercede.[35]

Although the mining office no longer had the right to establish the price of coal for each mine, it did together with the representatives determine a price which was to be the basis for tax computation. Although this was to be the sales price of coal, the mining office had no control over the actual market price. To establish the tax price, the mining office did have the right to inspect the company's books. Furthermore, the person responsible for bookkeeping and accounting had to take an oath as to their correctness.[36]

The law of 1851, its instructions, additional regulations and ministerial explanations made the acquisition of mining property considerably less restrictive. Prior to

1851 the complicated procedure had as its major objectives the control not only of the number of firms but also of the size of the mining operation. After 1851, the intention of the reforms was precisely the opposite, the encouragement of large scale mining. The internal reforms of the mining company had already facilitated consolidation, but the restriction on consolidations were further reduced. The reform of the procedure acquiring mining property focused on the searching for minerals, and satisfying the requirements for claiming a field.

The ministerial regulation of March 31, 1852,[37] abolished all older or conflicting regulations dealing with Schürfen[38] and Muten.[39] Searching for minerals required approval by the mining office of a written request describing in detail the area. The older regulations had prohibited one person or his relatives from acquiring several adjacent fields. Now this ruling was abolished. Furthermore, fields where permission to search had been received could also be transferred to other interested parties by written agreement.

According to the same regulation, each application for a claim made by a company had to include a list of its shareholders. The claim became legally effective if proof of the existence of the mineral had been furnished and accepted by the mining administration. If this had not occurred then the claim was again declared to be open to all interested parties.

For the first time paragraph 32 of the regulation mentioned the possibility that the result of drilling could satisfy the requirement for a claim.[40] However, the paragraph was vague, and the ministerial order of May 10, 1853,[41] and the ministerial regulation of November 2, 1853,[42] attempted to clarify the legal points and the intention of the paragraph in question. The result was that under certain circumstances drilling was to be accepted as sufficient evidence for a claim. A field so claimed gave the interested party the right to apply for a formal concession for the field in question.

Finally, in 1860 and 1861, two other important laws completed the major reforms and their objectives. The law of May 21, 1860,[43] dealt with the supervisory function of the mining administration as well as with the mine workers. Paragraph one of the law clearly stated that the mine owners were no longer under the supervision of the mining office in regard to the extraction and use of the minerals--the mining operation. The mining office did have the supervisory power where "it was necessary to maintain the effectiveness of mining, the safety of the operation, the surface, and the protection of the life and health of the workers."[44] Therefore, the technical plan of operation still had to be approved by the mining administration.

The 1860 law drastically changed the relationship between the mine owners and the mine workers, especially guild workers. Contractual freedom was established between employer and all employees. Now the mine workers could choose their own employers. The mining office was not allowed to take part in any hiring or dismissal of the mining personnel nor in the determination of their salaries and wages. But the law did maintain the right of the state to control the qualifications of the technical and business personnel.

The law of June 10, 1861, defined the function of the mining headquarters in Prussia.[45] Paragraph one abolished all regional mining offices under each mining headquarters, or *Oberbergamt*. Their previous jurisdiction was to be exercised directly by the *Oberbergamt* itself. The law reorganized Prussia into four mining areas, each with an *Oberbergamt*. In the case of the Ruhr region, the law abolished the mining offices at Bochum and Essen, but the *Oberbergamt* at Dortmund remained.

Of course, between 1851 and 1861 there were numerous other ministerial orders, regulations, and instructions. Most of these dealt with the application of the laws mentioned, others focused on the supervision by the mining administration of the technical aspect of a mining operation and the qualifications of the managerial personnel. Still others concentrated on the police power of the mining administration concerning the safety of the operation and protection of the life and health of the workers. Finally a great number of regulations and instructions focused on the safety of using machinery, especially steam engines.[46]

D. Minister Von der Heydt and the Opposition to the Reform Laws

Prior to 1851 the basic conflict and tension had been between the state and private enterprise. Both agreed that a general reform of the old mining law was necessary. They did not agree on the substance and particularly on the precise relationship between private enterprise and the state. Of course, the various interests of the provincial Diets and provincial mining administration further complicated the conflict. Because of the various interests, no general mining codification was accepted. Nevertheless, with the introduction of the individual reform laws in 1851, the tension between the government at Berlin and private mining enterprise had been eased. Private enterprise had won its major objectives, greatly reducing state restriction on the activities of the mining company and the mining operation.

After 1851 the pressure for continuing reform of the old mining law came from the state and its highest mining official, Minister of Commerce, August von der Heydt. Now the real tension and conflict was between the ministerial level and the mining administration, between the Minister himself, as a reformer, and the provincial mining headquarters, chiefly the <u>Oberbergamt</u> Dortmund. Basically the

Oberbergamt wanted to retain more authority than the minister allowed.

After 1848 August von der Heydt was one of the few liberal reformers in high government position. He had come from the Wupper Valley city of Elberfeld where he had received his training as a merchant and banker in his father's banking firm, Gebrüder Kersten.[47] Since the late 1820's he had been actively engaged in local, provincial, and finally national politics. In November of the revolutionary year 1848 he was appointed Minister of Commerce. Until his dismissal, after 14 years, and despite his liberal attitude, he weathered the changing Prussian politics, including the change in ministry from Manteuffel to Hohenzollern von Auerswald in 1858. It is interesting that in the same year a group of Rhenish industrialists had even appealed to the Prince-Regent (later king Wilhelm) to keep von der Heydt in his position because he had so well served the interests of the Rhenish and Westphalian provinces. In 1862, he unfortunately became entangled in the military budget conflict and shortly thereafter was dismissed.

The major reason for his long tenure in office was his unquestionable organizational talents and ample accomplishment, many directly benefitting the state. During his time in office von der Heydt was constantly organizing, promoting, and reforming. Although before 1848 he had already promoted and participated in railway construction as a private entrepreneur, at the same time he had advocated state involvement. As minister he successfully promoted state ownership. He had argued that the financial burden for private enterprise was too great and in addition such an important new means of public transportation should be state owned or controlled. During his tenure he also developed and organized the postal and telegraph system, urged improvement of roads, rivers, and the construction of canals, encouraged the introduction of new methods in the linen industry in Westphalia and Silesia, and accomplished the total reorganizations of trade and technical schools.

While von der Heydt believed in state ownership or control of public transportation, communication, and education, he was also a firm believer in the benefits of private enterprise. Consequently he was actively engaged in the abolition of old restrictions imposed on private enterprise. This was particularly the case in the mining industry. He, more than anyone else, should be credited for the relatively quick and successful reform of the _Direktionsprinzip_, especially in the Ruhr mining region. It is interesting that while one known liberal reformer, Freiherr vom Stein, had successfully installed the _Direktionsprinzip_ more than 60 years earlier, another successfully abolished it.

While Minister von der Heydt impatiently pushed for reform and the immediate and effective application of the Reform Law of 1851, the mining administration, especially in the Ruhr region, resisted and only reluctantly applied the law. Therefore, after 1851 a major objective of the reform movement was not only to reform the old legal system but, equally important, to reform the mining administration itself. After all, the administration of the *Direktionsprinzip* had existed for nearly ninety years. The mining bureaucracy had had ample time to develop its own tradition, loyalties, and even philosophy. Now the reform laws required a restructuring of the administration and a new relationship to private enterprise. The reform laws demanded a totally new attitude from the mining administration. They imposed on the mining offices a shift from the direct exercise of authority to supervision, from orders to advice, from direction to inspection. To many bureaucrats, this was a virtual revolution and they resisted it.

The major conflict between the Minister of Commerce and the resisting *Oberbergamt* Dortmund concerned the procedures for claiming a field and granting a concession and the qualification of the technical and business personnel employed by the private company.

The ministerial order of May 10, 1853, had changed the basis for granting a claim.[48] The old law had required that the mineral be proved worth mining and that it be literally uncovered to eyesight in order to assure a cost recovering operation. The 1853 order abolished these requirements; all that had to be proven was the existence of the mineral.[49]

The order intended to liberalize this requirement even further. An interested party did not necessarily have to prove the existence of the mineral in question where it was clear from evidence such as neighboring fields. Unfortunately, the order did not clarify the question whether the results of drilling were acceptable as proof for the existence of minerals.

The regulation of November 2, 1853, answered a question from the mining headquarters, "if concession can be granted after a claim had only been proven by the results of drilling holes."[50] All but one Oberbergamt, the Oberbergamt Dortmund, had answered their own question in the affirmative.[51] In the regulation the Minister restated the prerequisites for acquiring a field concession. "The interested party must as before uncover to eyesight the mineral involved,"[52] however, "in special cases," where other conditions indicate the existence of coal, "drillings also should be accepted."[53] These drillings had to be undertaken in the constant presence of a mining official.

The Oberbergamt Dortmund objected to drilling as sufficient evidence for either a claim or a concession, and questioned the legality on the basis of article III of the Cleve-Mark Mining Codification of 1766.[54] In the above regulation of 1853, Heydt replied that this part of the law had been unclear as to whether the uncovering of the coal had to be done prior to the granting of a field concession. As additional support he cited other later mining codifications including Silesia's in 1769 and the mining part in the <u>Landrecht</u> which indicated that a concession could be granted before the coal was uncovered. Further Heydt argued that the new circumstances in mining could not possibly have been foreseen by the old law. It did not even mention shaft mining. Therefore, even when applying the old law to new situations, a liberal interpretation becomes absolutely necessary.

Under the <u>Direktionsprinzip</u> in the Ruhr region a shaft had been necessary to prove the existence of coal before a field concession was granted. To this Heydt advanced the argument long voiced by shaft mine owners, that the risk and the cost of a shaft, without the certainty of the field concession, was far too great and unreasonable. It suffocated private initiative.

Therefore, a field concession could now be acquired on the basis of the results of drillings, only without the cost of constructing a shaft. Furthermore, enough geological knowledge of the Ruhr coal field had accumulated so that the "special case" became general practice by the late 1850's. Finally, it is obvious that all the procedures necessary for acquiring mining property were now essentially reduced to proving of the existence of the mineral in any reasonable manner. The method was irrelevant. Minister Heydt's intention had been to allow private enterprise to make the relevant decisions, such as if the mineral was worthy of mining.

The other major conflict between Minister von der Heydt and the Dortmund mining office centered on the reluctance by the mining office to hand over the management of the mining operation to the private mining company. This transfer of control was of course precisely the purpose of the 1851 law, but the same law had left the mining office in control of the qualifications of the technical and business personnel employed by the mining company. On the surface the conflict was a disagreement as to the necessary qualifications of these personnel. Several ministerial instructions were specifically directed to the Oberbergamt,[55] but it is already clear from an instruction of May 30, 1852, that the Oberbergamt was attempting to misuse its reduced authority by demanding higher qualification examinations than had been the practice under the Cleve-Mark Mining Codification.[56]

Minister von der Heydt replied to the mining office that the qualifications for the managerial personnel were to be "practical education and experience, and little theoretical knowledge was necessary."[57] The Minister warned the Oberbergamt not to introduce these examinations, demanding the same qualification from the managerial personnel that the district officials possessed. He reminded the mining office that there were no such examination for its own district officials. Introduction of such an examination would certainly be opposed, since "there already exist a sufficient number of formally trained and examined personnel for all such positions."[58]

The major reason for this attempt by the <u>Oberbergamt</u> had been to secure jobs for its own officials. The 1851 law endangered the jobs of a large number of the lower echelon mining officials. Von der Heydt had attempted to soothe some of the immediate fears of the mining office. He pointed out that it could be expected that the mining companies of larger mining operations would shortly employ their own technical directors. (This had already occurred even in the Ruhr region.) But it would take some time until the smaller companies could consolidate and afford such a technical management. Therefore, "the district officials under whose jurisdiction these mines are, will for some time continue to be responsible for the supervision of their operations."[59] Of course they would not have the same authority as before but rather would be limited to the supervision of the more technical and specialized aspects of the operation. In any case such help was not to be withheld from mine owners if they requested it.

Minister von der Heydt clearly stated the objectives of the reform law of May 12, 1851, in the instruction of March 6, 1852 was to:

> ...provide a transition between the previously state directed management and the future self-management of the private mining company. During such a transition period, no general standards can be established.[60]

To the report by the <u>Oberbergamt</u> Dortmund requesting higher qualification examinations, the minister replied:

> Measures which are requested in this report would make the transfer of management to the mine owners so difficult that the present situation would continue for a long time. But this would be in absolute opposition to the meaning and purpose of the previously mentioned law /1851/.[61]

Von der Heydt reprimanded the Oberbergamt's attitude and instructed it to apply the law and its instruction.

An instruction of February 15, 1853, concerning the transfer of management, indicated that the Oberbergamt Dortmund was still reluctant to carry out the 1851 law. Heydt emphasized that the transfer must be undertaken despite the objections of some mine owners. "There has now been sufficient time to effect the transfer gradually."[62] Furthermore the tax law of 1851 "had reduced taxes more than enough to cover the cost of self-management."[63] According to Heydt the supervisional tax of one per cent was insufficient to cover the cost of the mining administration. This cost could only be reduced by withdrawing the district officials and forcing the companies to manage their own operations. Again Heydt emphasized that the purpose of the 1851 law had been to create the necessary legal framework for the transfer of management.

Apparently the Oberbergamt Dortmund was also reluctant to allow the operation of foreign firms in the Ruhr region. The law of May 4, 1846 had permitted foreign corporations to acquire property but with the direct consent of the state.[64] In the order of November 11, 1853, the minister reprimanded the mining office and reminded it that "foreign corporations or other legal organizations desiring a field concession should be granted them, after they had satisfied the other regular requirements."[65]

As late as 1856 the Minister again instructed the Oberbergamt that "when difficulties arise in the administration of the law of May 12, 1851, they should be solved in an imaginative, benevolent, and considerate manner using administrative methods."[66] He implied that special cases should be solved without constant ministerial directions.

All of von der Heydt's instructions and explanations to the various mining offices urged the transfer of authority to the private mining company. This transfer was to be made as smoothly but also as speedily as possible and the mining administration was to help with its personnel or technical advice. In no case should the mining administration hinder or oppose such a transfer. Finally, the Minister repeatedly emphasized, that it was not the letter of the 1851 law but rather the spirit that should dominate its application.[67]

FOOTNOTES - CHAPTER V

1. Dietrich Eichholtz, _Junker und Bourgoisie_. Specifically discusses the various interest groups and their effect on the state's railway policy. Werner Sombart, _Die deutsch Volkswirtschaft im 19. Jahrhundert und im Anfang des 20. Jahrhunderts_, (Stuttgart, 1954).

2. Werner Sombart, _Wirtschaftsgeschichte_..., pp. 63-65.

3. _Ibid_., pp. 65-66.

4. Agnes Prym, _Staats wirtschaft_..., p. 22.

5. W.J.D. Brassert, "Das Recht des Mitbaues zur Hälfte," _Z.f.B.H.S.-W._, IV, 23.

6. Sombart, p. 65.

7. Otto Paulus, _Die Stellung des Rheinlandes zur Schutzidee in der Preussischen Wirtschafts-und Sozialpolitik von 1815 bis um die Mitte des 19. Jahrhunderts_ (Diss. Köln, 1927) pp. 7, 8; Spethmann, _Das kuhrgebiet_, vol. II, pp. 173-178.

8. Spethmann, p. 173.

9. _Ibid_., pp. 177, 178; Reuss, pp. 63-65. Lists the installation of 9 steam engines, 1808-1814.

10. Heinrich Achenbach, _Das französische Bergrecht und die Fortbildung desselben durch das preussische Berggesetz_, (Bonn, 1869). He provides a history of the French mining law from April 28, 1791 to April 10, 1810 and compares the French Law to the Prussian mining law indicating influences of the French on the Prussian law.

11. Achenbach, p. 90.

12. Achenbach, "Eigeninteresse als Bürge des öffentlichen Interesse im französischen Berggesetz von 1810," _Z.f.Bergrecht_, III (1863), 82.

13. Boldt, _Staat und Bergbau_, p. 6.

14. Boldt, p. 6; Brassert, "Die Bergrechtsreform in Preussen," Z.f.Bergrecht, II (1862), 234-253. The rest of this section is essentially based on Brassert's history of the mining revision attempts 1826-1851. Brassert was editor of the only Prussian Mining Law journal, Z.f.Bergrecht, and was the father of the General Codification of 1865.

15. O.B.A. Dortmund, "Gutachten verschiedener Beamten über den Entwurf des neuen Berggesetzes, 1837, und Sitzungsproto kolle der zur gutachtlichen Vorberatung eines neuen Berggesetzes berufenen Commission," (Druckschrift, O.B.A. Dortmund, B.Gr. Nr. 50a).

16. Carnall, "Übersicht, contains in a rough form the December 1848 revision. He was one of the leaders in this commission's work.

17. Z.f.B.H.S.-W., I, 15.

18. This report was given by the Berichterstatter f.d. Haus der Abgeordneten, and subsequently published as "Technische Bürokratie und Unternehmer im Prinzipienstreit um das Berggesetz," Z.f.Bergrecht, 1865, 293.

19. "Gesetz über die Verhältnisse der Miteigenthümer eines Bergwerks, 12. Mai, 1851," Z.f.B.H.S.-W., I (1854), 24-28; hereafter cited as "Shareholder Law."

20. "Gesetz über die Besteuerung der Bergwerke, 12. Mai 1851", Z.f.B.H.S.-W., I (1854) 15-17; hereafter cited as Tax Law.

21. "Gesetz über die Aufsicht der Bergbehörden über den Bergbau und das Verhältnis der Berg-und Hüttenbetriebe, 21. Mai 1860," VIII, 217-220 Z.f.B.H.S-W. Hereafter cited as Law of May, 1860.

22. "Allgemeines Preussisches Berggesetz, 24. Juni 1865," Z.f.B.H.S.-W., XIII, 29-65.

23. "Instruction to the tax law," Z.f.B.H.S.-W., I, 17-21.

24. Law of May, 22, 1861, Z.f.B.H.S.-W., IX, 206, reduces taxes to 4 per cent effective Jan, 1862. Law of October 20, 1862, vol. X, 240, taxes reduced to 3 per cent effective Jan. 1863. Law of June 17, 1863, XI, reduced taxes to 2 per cent effective Jan. 1864 and to 1 per cent effective Jan. 1865.

25. Shareholder Law, SS 2-9.

26. Ibid., SS 2-7.

27. Ibid., SS 14-15.

28. Ibid., SS 18, lists 17 responsibilities of a representative or board.

29. March 6, 1852, Instruction to the Shareholder Law, Vol. I, 28-32.

30. May 22, 1852, Instruction to the Shareholder Law, Vol. I, 36.

31. Instruction of March 6, 1852, p. 29.

32. Ibid., p. 29.

33. Ibid., p. 30.

34. Ibid., p. 31.

35. Ibid., p. 31.

36. Instruction, March 6, 1852, p. 31.

37. "Circular Verfügung, 31. März 1852, betreffend die Ausfertigung der Schürferlaubnis..." Z.f.B.H.S.-W., I, 41-47.

38. Ibid., "schürfen," SS 1-18.

39. Ibid., "muthen", SS 19-44.

40. Ibid., SS 32.

41. Ministerial Order of May 10, 1853, I, 122.

42. Ministerial Regulation of Nov. 2, 1853, I, 223-225.

43. Law of May 21, 1860, VIII, 217-220.

44. Ibid., SS 1.

45. Law of June 10, 1861, IX, 222-224.

46. Regulation of June 18, 1855, and the Law of May 7, 1855, IV. Both deal with the safe operation of steam engines.

47. A. Bergengrun, Staatsminister August von der Heydt, Berlin: 1908.

48. Ministerial Order of May 10, 1853, I, 122.

49. Ibid.

50. Regulation of Nov. 2, 1853, I, 223-224.

51. Ibid., 223.

52. Ibid.

53. Ibid.

54. Ibid., 224.

55. Instruction to the OBA Dortmund, May 30, 1852, I, 36-38; Instruction to the OBA Dortmund, Feb. 15, 1853, I, 39-40; Instruction to the OBA Dortmund, April 2, 1856, IV, 50, 51.

56. Ibid., Instruction, May 30, 1852, I, 37.

57. Ibid., p. 36.

58. Ibid., p. 37.

59. Ibid., p. 36.

60. Ibid., p. 37.

61. Ibid., p. 37.

62. Instruction, Feb. 15, 1853, p. 39.

63. Ibid.

64. "Law of May 4, 1846," Z.f.B.H.S.-W., II, 126.

65. Ibid.

66. Instruction of April 2, 1856, IV, 50, 51.

67. Ibid., 51.

CHAPTER VI

THE REFORMED DIREKTIONSPRINZIP AND ECONOMIC CHANGE IN THE MINING INDUSTRY: 1851-1865

The previous chapter described the modification of the *Direktionsprinzip* by the state after 1851. This chapter discusses the response of private enterprise to the reform and to the general economic changes in the Ruhr between 1851 and 1865. The first section analyzes the general growth of the Ruhr coal industry and large mining operations. The second section discusses the impact of the tax and shareholder laws on the industry and concentrates on the spread of joint-stock companies as the dominant business and financial organization appropriate for large-scale mining operations. The last section deals with the industry's response to changes in the market.

A. The Growth of the Ruhr Coal Industry and Large Operations, 1851-1865

The reform laws for the first time allowed private enterprise a range of economic decisions which had not been possible under the *Direktionsprinzip* before 1851. The laws of 1851, their subsequent instructions, and additional ministerial orders were designed to encourage private initiative and enterprise by removing major restrictions from private entrepreneurial activities. This development was further encouraged by the general upswing in economic activities in the 1850's.

While the reform had been timely for the direct participation by private enterprise in the economic boom, the Ruhr mining industry also experienced the full impact of a major depression after 1857. In both instances the industry did remarkably well. Private enterprise was able to respond to the rapidly changing economic conditions. Tables 6.0 through 6.12 at the end of this section provide the general statistical data for the Ruhr coal industry between 1851 and 1865.

During the entire period between 1851 and 1865 coal production increased constantly with the exception of a slight drop in 1859. The annual growth rate for the entire 100 years of the Direktionsprinzip was little more than 5 per cent, but the average annual rate for the reform period was 12.3 per cent. Total production increased from 1.7 million tons in 1850 to 9.2 million tons in 1865. Production doubled in the five years after 1850, an unprecedented occurrance; in 1865 production was more than 5 1/2 times the 1850 level (Table 6.2). The average mining operation in 1850 had an output of 8,400 tons and employed 64 workers. In 1865 the average mine produced nearly 40,000 tons and employed 184 workers. While the average output per worker was 131 tons in 1850, it had increased to 215 tons in 1865 (Table 6.1).

Tables 6.3 and 6.4 clearly indicate the boom and depression periods. The largest boom was in the three years between 1854 and 1856. It was associated with an equal rate of increase in production and employment, 18 per cent, and also with the highest yearly price increase, 11 per cent, as well as the highest yearly increase in the number of mines, 13.7 per cent. The depression started in late 1857 when prices began to decrease. During the worst of the depression years, 1860-1862, the annual rate of price change was minus 14.6 per cent; the number of mines also decreased at 3.2 per cent a year; eliminating many of the small scale marginal firms producing less than 30,000 tons (Table 6.5). Yet most significantly the annual growth rate of production was 17 per cent or nearly equal that of the boom period. The mining industry was successfully adjusting to the adverse economic conditions.

In general until 1858, the industry responded to increases in demand by increasing employment and number of mines. As Table 6.1 indicates, average productivity per worker actually declined for a few years. The continued increase of production during the depression is accounted for by the entrance of new and more efficient large scale operations. These mines entering production during the depression had been formed during the boom period, and were mostly joint-stock companies. The effect of these new operations was not only increased production but also further decreased prices.

While in the early 1850's the largest mines had an output between 50,000 and 100,000 tons, the typical new operation by the late 1850's exceeded 100,000 tons, employing 500 to 600 workers. Even during the depression year of 1860 three mines, each with an output of over 130,000 tons, entered production. In 1864, there were nine mines producing more than 130,000 tons, four of these producing more than 200,000 tons. In 1850 the eight largest operations, each with an output over 50,000 tons, shared 30 per cent of the total Ruhr production. In 1864 the 55 operations with an output over 50,000 tons shared 67 per cent of the total production, but the eight largest operations had a combined output equivalent to the entire industry's production in 1850.

In general the new firms in the 1850's operated on fields the size of at least three old fields. They not only used steam engines for pumping water and lifting coal, but by the late 1850's had installed other machinery making coal production more efficient. By the late 1850's many of the large operations added coal washing, sorting, and coking operations.

Between 1850 and 1865 the largest operations were clearly more efficient in terms of productivity per worker. The largest operation in each year between 1850 and 1865 had the highest productivity. Furthermore, the average productivity was also greater in the mines with an output over 50,000 tons than the average of the industry as a whole. In all three categories the productivity per work increased during the entire period. Finally, while there was a large difference between the three categories in the early 1850's, by 1864 the productivity per worker of the largest mine, the mines with an output over 50,000 tons, and the entire industry had gradually converged to the same level (Table 6.8). The large and more efficient operation increasingly dominated the production and nearly all the firms in the industry had gradually adopted similar methods, such as shaft mining, and similar technology, such as steam engines. The use of machines, particularly steam engines, increased rapidly, as did their horse power. In 1850 there were 123 steam engines engaged in the Ruhr coal industry; in 1865 their number had increased to 462. In the Essen region, where most of the large shaft mines were located, the horsepower per mine continuously exceeded the average for the industry. Meanwhile the average horse power per mine more than doubled from 53 in 1850 to 120 horsepower per mine in 1865 (Table 6.9).

B. The Effect of the Reform and the Growth of the Joint-stock Company

1. The tax reform

Heinrich von Achenbach, a contemporary high mining

TABLE 6.0

PRODUCTION, EMPLOYMENT, PRICE, MINES: 1850-1865

Year	Number Mines	Total Employment (100)	Total Production (1000 tons)	Average Price per ton (R.M.)
1850	198	12 7	1,666	5.93
1851	189	14 3	1,804	5.87
1852	178	15 2	1,956	5.33
1853	194	17 2	2,187	5.56
1854	205	20 0	2,719	6.79
1855	240	23 8	3,317	8.32
1856	282	28 5	3,575	8.99
1857	299	30 3	3,725	8.91
1858	292	32 7	4,006	8.84
1859	286	30 1	3,889	7.87
1860	281	29 3	4,366	6.44
1861	275	31 5	5,555	5.13
1862	266	32 9	6,242	4.81
1863	241	33 4	6,875	4.36
1864	232	38 7	8,146	4.55
1865	234	43 1	9,277	5.02

Source: Reuss: <u>Mitteilungen</u>, p. 84

TABLE 6.1

AVERAGE PRODUCTION AND EMPLOYMENT PER MINE AND PRODUCTIVITY PER WORK

Year	Average Production Per Mine	Average Production Per Work	Average Employment Per Mine
1850	8,414	131	64
1851	9,545	126	76
1852	10,988	129	85
1853	11,278	127	89
1854	13,265	136	98
1855	13,821	139	99
1856	12,671	125	101
1857	12,458	123	101
1858	13,719	123	112
1859	13,598	129	105
1860	15,537	149	104
1861	20,200	176	115
1862	23,466	190	124
1863	28,526	206	139
1864	35,112	210	167
1865	39,645	215	184

Source: Derived from Table 6.0

TABLE 6.2

INDEX NUMBERS OF PRODUCTION, MINES, EMPLOYMENT AND PRICE
(1850 = 100)

Year	Production	Employment	Price	Number Mines
1850	100	100	100	100
1851	108	113	99	95
1852	117	120	90	89
1853	131	135	94	97
1854	162	157	114	103
1855	200	187	140	120
1856	214	224	151	141
1857	223	238	150	150
1858	240	257	149	146
1859	239	237	133	143
1860	262	230	108	141
1861	332	248	87	138
1862	374	258	81	133
1863	412	263	73	121
1864	488	300	77	116
1865	555	338	84	117

Source: Derived from Table 6.0

TABLE 6.3

YEARLY PERCENTAGES CHANGES, 1850-1865

Year	Production	Price	Employment	Mines
1850	+20	- 3	+ 5	- 5
1851	+ 8	- 1	+13	- 5
1852	+ 8	- 9	+ 6	- 6
1853	+12	+ 4	+13	+ 9
1854	+24	+ 7	+16	+ 6
1855	+22	+18	+19	+17
1856	+ 8	+ 8	+19	+18
1857	+ 4	- 1	+ 6	+ 6
1858	+ 8	- 1	+ 8	- 2
1859	- 3	-11	- 9	- 2
1860	+12	-18	- 3	- 2
1861	+27	-20	+ 8	- 2
1862	+12	- 6	+ 5	- 3
1863	+10	- 9	+ 1	-10
1864	+19	+ 4	+16	- 4
1865	+14	+ 9	+11	+ 1

Source: Derived from Table 6.0

TABLE 6.4

THREE YEAR AVERAGE RATES OF CHANGE: 1851-1865

Years	Production	Employment	Mines	Price
1851-53	+ 9.3	+10.7	- 0.3	- 2.0
1854-56	+18.0	+18.0	+13.7	+11.0
1857-59	+ 3.0	+ 1.7	+ 0.7	- 4.3
1860-62	+17.0	+ 3.3	- 2.3	-14.6
1863-65	+14.3	+ 9.3	- 4.3	+ 1.3

Source: Derived from Table 6.0

TABLE 6.5

NUMBER OF MINES AND SIZE CLASSES

Size class of mines by output (* 999 or +1000 tons)	1850°	1852	1853	1854	1855	1856	1857	1858	1859	1860	1861	1862	1863	1864
All Mines	198	178	154	205	240	282	299	292	286	281	275	266	241	232
0- 29*	178	155	129	174	202	240	257	269	242	232	221	210	170	156
30- 39	6	9	12	12	11	10	12	8	16	16	13	9	18	8
40- 49	6	4	2	4	6	7	7	9	7	10	12	12	7	12
50- 59	5	8	5	2	3	2	5	7	2	2	10	9	10	5
60- 69	1	2	2	3	3	6	3	4	4	4	6	5	9	12
70- 79	1		4	4	4	1	3	2	3	6	2	7	5	7
80- 89	1		1	4	5	7	4	2	3	2	2	3	6	7
90- 99				2	3	3	4	3	4	2	4	3	2	4
100- 09					-	1	1	3	2	1	-	2	2	7
110-119					2	1	2	2	1	2	-	1	3	2
120-129					1	2	-	1	2	1	1	-	-	2
130-139							1	2		-	2	-	-	1
140-149							1			1	-	2	1	1
150-159										1	1	1	-	2
160-169											1	1	3	1
170-179											-	1	-	-
180-189											1	-	-	-
190-199											-	1	1	-
200-209											1	1	1	-
above 210												1	2	4

°Data for 1851 not available

Source: Zeitschrift für Berg-, Hütten-, Salinenwesen, Vols. I-XIII

TABLE 6.6

PRODUCTION OF MINES OVER 50,000 TONS

Year	Total Number Mines	Number Mines Over 50,000 tons	Percentage Production Mines over 50,000 tons
1850	198	8	30
1851	189	-	-
1852	178	11	31
1853	194	11	33
1854	205	15	41
1855	240	21	52
1856	282	24	53
1857	299	24	54
1858	292	26	60
1859	286	21	47
1860	281	23	48
1861	275	31	48
1862	266	38	58
1863	241	45	62
1864	232	55	67
1865	234	-	-

Source: Zeitschrift für Berg-, Hütten-, Salinenwesen, Vols. I-XIII and Reuss, "Mitteilungen...", Festschrift, p. 85.

TABLE 6.7

MINES PRODUCING OVER 50,000 TONS

Size Class (1,000 tons)	Number of Mines 1852	1856	1860	1864
Over 200	0	0	0	4
Over 150	0	0	1	7
Over 130	0	0	2	9
Over 100	0	4	6	20
Over 80	0	14	10	31
Over 60	2	22	21	50
Over 50	10	24	23	55

Source: Derived from Table 6.5

TABLE 6.8

OUTPUT PER WORKER, 1850-1865

Years	Production per worker in largest mine (tons)	Average Production per worker in mines producing over 50,000 tons	Total Average Production per worker (tons)
1851	236	195	131
1852	-	-	126
1853	-	-	129
1854	201	169	127
1855	218	173	136
1856	233	174	139
1857	194	162	125
1858	194	160	123
1859	190	167	123
1860	190	169	129
1861	205	187	149
1862	219	189	176
1863	233	200	190
1864	235	216	206
1864	224	215	210
1865	-	-	215

Source: Z.f.B.H.S.wesen, Vol. I-XIII.

TABLE 6.9

STEAM ENGINES AND THEIR HORSE POWER
IN THE ESSEN, BOCHUM REGIONS

	Number of steam engines				Horsepower per mine		
Year	Bochum	Essen	Total	Horse-power	Bochum	Essen	Total
1834	-	-	50	-	-	-	-
1843	-	-	95	-	-	-	-
1850	-	-	123	-	-	-	-
1851	-	-	142	9,860	-	-	53
1852	72	85	157	11,570	52	97	65
1853	78	88	166	13,030	49	133	67
1854	86	80	166	13,110	50	97	64
1855	101	88	189	14,520	52	80	60
1856	145	96	241	17,670	59	75	63
1857	179	122	301	21,460	65	85	72
1858	203	140	343	25,450	78	107	87
1859	225	147	372	27,170	83	120	95
1860	226	135	361	30,770	97	134	109
1861			394	32,880			120
1862			430				
1863			397				
1864			414				
1865			462				

Source: Z.f.B.H.S.-Wesen, Vols. I-XIII, and Reuss, "Mitteilungen...," Festschrift, p. 85.

TABLE 6.10 *

PRODUCTION AND PRICE: 1788-1803, 1835-1850, 1850-1865

Production			Average Price		
(1)	(2)	(3)	(4)	(5)	(6)
100	100	100	-	100	100
103	95	108	-	94	99
107	111	117	-	93	90
103	127	131	-	103	94
138	130	162	100	109	114
141	126	200	108	103	140
123	139	214	111	109	151
127	144	223	108	113	150
153	137	240	111	112	149
157	152	239	111	107	133
164	161	262	112	105	108
178	172	332	121	103	87
180	184	374	121	108	81
157	170	412	222	100	73
163	176	488	144	101	77
157	212	555	137	98	84

(1) 1788-1803, base year 1788 = 100 per cent

(2) 1835-1850, base year 1835 = 100 per cent

(3) 1850-1865, base year 1850 = 100 per cent

(4) Base Year 1792 = 100 per cent

(5) Base Year 1835 = 100 per cent

(6) Base Year 1850 = 100 per cent

Source: Derived from Tables 2.1, 2.2, 6.0.

*Tables 6.10, 6.11, and 6.12 compare the growth of the coal industry in three 16 year periods. The first 16 years, 1788-1803 experienced the full implementation of the <u>Direktionsprinzip</u> and the state's active role in the mining industry and the region. The 16 years 1835-1850 are the last years of the state control system. The following 16 years, 1850-1865 correspond to the period of reform of the <u>Direktionsprinzip</u>. The regulation of the price had been a

Table 6.10 - continued

major objective of the mining administration before 1850. While the coal prices were increasing between 1792 and 1803, they remained virtually constant from 1835 to 1850. The prices in the following 16 years fluctuated sharply, more clearly reflecting the market condition.

TABLE 6.11*

MINES AND EMPLOYMENT: 1788-1803, 1835-1850, 1850-1865

Mines			Employment		
(1)	(2)	(3)	(4)	(5)	(6)
100	100	100	100	100	100
108	105	95	100	116	113
98	110	89	100	127	120
93	110	97	100	137	137
91	112	103	117	140	157
94	116	120	125	153	187
92	112	141	108	156	224
86	118	150	117	159	238
91	111	146	125	159	257
88	113	143	125	171	237
91	113	141	142	177	230
88	115	138	142	185	248
93	118	133	125	201	258
93	116	121	108	195	263
94	109	116	117	205	300
82	105	117	133	215	338

(1) Base Year 1788

(2) Base Year 1835

(3) Base Year 1850

(4) Base Year 1788

(5) Base Year 1835

(6) Base Year 1850

Source: Derived from Tables 2.1, 2.2, 6.0.

*Before 1850, another major objective under the Direktionsprinzip had been to control the number of mining operations (see Table 6.11). Between 1788 and 1803 the number of mines decreased showing the mining administration's effort to eliminate the most inefficient mines. Between 1835 and 1850 the number of mines remained constant, indicating the mining administration's success in controlling the number of mines. Again the following 16 years show the fluctuation in the number of mines; the administration had little control over the entrance and exit of firms in the industry.

TABLE 6.12

OUTPUT AND EMPLOYMENT PER MINE, OUTPUT PER WORKER: 1788-1803, 1835-1850, 1850-1865

Output per Mine			Employment per Mine			Output per Worker		
(1)	(2)	(3)	(4)	(5)	(6)	(7)	(8)	(9)
100	100	100	100	100	100	100	100	100
95	91	114	100	110	118	103	83	96
109	101	131	100	116	133	107	87	98
195	115	135	114	126	139	103	92	97
153	114	158	129	126	153	118	91	104
152	108	168	129	132	155	113	83	106
135	124	151	114	139	158	115	90	96
147	121	150	143	136	158	109	90	94
169	124	163	143	145	175	122	86	94
178	135	162	143	151	164	126	89	98
178	142	185	157	158	163	113	90	114
203	149	240	157	161	179	126	92	134
194	154	280	129	171	194	144	91	145
169	147	339	114	165	217	144	87	157
175	162	418	129	187	261	140	86	160
194	204	471	157	206	287	119	99	164

(1) Base Year 1788

(2) Base Year 1835

(3) Base Year 1850

(4) Base Year 1788

(5) Base Year 1835

(6) Base Year 1850

(7) Base Year 1788

(8) Base Year 1835

(9) Base Year 1850

Source: Derived from Tables 2.2, 6.1.

official pointed out, "the shareholder law rid the mining administration of its private functions, transforming it into a state agency with only police and supervisional powers;" the companion tax law "reduced the tenth to the twentieth and introduced a supervisory tax giving this new tax the character of a general business tax."[1] As discussed previously the taxes were felt as a major economic burden particularly by large-scale shaft mining operations since the 1830's. These leading mining interests considered tax reform essential for the continuous growth of the industry. Consequently one of the first reforms was the tax law which together with the shareholder law of the same day introduced the reform. The tax law by reducing and simplifying the taxes made them similar to a normal state tax paid by other industries and businesses.

As Table 6.13 indicates, the tax law lowered the state mining income from 1,275,000 thalers or 13.0 per cent of the value of production in 1850 to 578,000 thalers or 5.7 per cent in 1852. While in 1850 the tax revenue from the tenth had been 821,000 and other special mining taxes (24 all together) amounted to 455,000 thalers, in 1852 state revenue from the twentieth was 473,000 thalers and from other taxes, mainly the one per cent supervisional tax, only 106,000 thalers. This sizeable tax reduction of 700,000 thalers financially benefitted the entire industry but especially the larger operations which had most heavily felt the tax burden. Finally, although the revenue dropped to 5.7 per cent in 1852 and was only 5.9 per cent of the value of production in 1855, the total tax income by 1855 surpassed that of 1850. This was basically due to the large production and price increases during the same years.

Despite the tax decreases to less than 6 per cent of the value of production, on the Left Bank of the Rhine, where the French code of 1810 was still in force, total taxes amounted to only 2.3 per cent of gross income.[2] Consequently, this provided an argument for future tax reduction. As the accompanying table indicates, between 1862 and 1865 the taxes were further reduced, finally amounting after 1865 to around 2 per cent of gross income.

2. The Economic Effects of the Shareholder Law

The shareholder law handed over management to private enterprise, encouraged large-scale mining through consolidation, and most important, it allowed free range to competition. A major purpose of the shareholder law was to create responsible private management for the firms to whom the mining operation could be entrusted. Despite the earlier complaints by mine owners about the mining administration's management of their operations only a relatively few firms responded immediately to the new law. By the end of 1852 only 15 operations in Essen and 15 in the Bochum region had taken over their own business and technical management. But

TABLE 6.13

TAXES: 1850-1866

Year	The 'Tenth' or reduced (1000 Marks)	Other Taxes (1000)	Total Taxes (1000)	Total Value of Sales (1000)	Percentage of value of Sales
1850	821	455	1,275		13.0
1851	(411) To June 1851	(225)	(635)		
1851	(235) After June 1851	(50)	(285)		
1851	(822) Projected	(449	(1,270)		13.0 Projected
1851	646 Actual	279	921		9.2 Actual
1852	473	106	578		5.7
1853	544	121	668		5.8
1854	835	183	1,018		5.9
1855	1,227	269	1,496		5.9
1856	1,389	306	1,696		5.9
1857	1,447	320	1,765		6.1
1858	1,545	330	1,884		5.8
1859	1,303	287	1,590		5.7
1860	1,188	268	1,456		5.6
1861	1,167	267	1,431		5.6
1862	1,002	388	1,290		4.7
1863	781	285	1,067		3.8
1864	614	338	952		2.7
1865	407	407	814		1.9
1866	-	-	1,008		2.2

Source: *Zeitschrift für Berg-, Hütten-, Salinenwesen*, Vo. XII, pp. 267-275; Vols. XIII, XIV.

nearly all 30 were shaft mines and 17 of these were the largest operations in the Ruhr region, producing more than half of the industry's output.[3]

The rest of the mines were reluctant to take over their own management, however. For example, of the 125 mines in Bochum,[4] 36 had expressed a desire to take over the entire management but only 15 actually did. Sixty-two mines wanted just the technical and bookkeeping management but not the business management, which was responsible for the general operation of the mine, including finances and sales. Twenty-two mines wanted further mining administration management.

Opponents of the reform laws both among private enterprise and the mining administration based their arguments on the traditional economic philosophy of the _Direktionsprinzip_ before 1851.[5] Competition would be disastrous to the mining industry, driving out many firms just for the sake of quick profit by a few; it would lead to increases in prices, hurting the consumer and thereby the market for the entire industry. The argument against the tax law was that the reduced taxes would only benefit the mine owners and not the consumer: it would lead to increased speculation and a price rise. Of course some of these arguments were based on false assumptions.

The shareholder and tax laws did not increase prices. Rather "for all mining products the prices had actually decreased, particularly the coal prices in 1852."[6] When the prices did increase in 1853 and continued to increase for the next 4 years, it was for different reasons. During this boom period, the increases in production and sales of coal was caused partly by the increasing use of railways for the transportation of the coal. But more important was the increasing use of coal in the metal producing industries, especially in the production and processing of iron. During these years there was a strong demand for the more expensive coal types such as coking coal which could only be produced by shaft mines.[7] In addition, the increasing prices of almost all competing fuels, especially wood and imported coal, benefitted the coal industry. In general, however, the entire period 1851-1865 was one of decreasing prices.

The shareholder law and the resulting competition did drive out the small and inefficient mines. Consolidation and formation of large mining operations did take place. The shareholder law, by forcing self-management on private enterprise, had the effect of reducing the number of firms in the industry. This in part explains the decrease in the number of mines in 1852. Many of the small and splintered operations could not afford a special manager and succumbed. Before 1851 the <u>Direktionsprinzip</u> and its administration actually protected the small and less efficient mining operations.

The shareholder law had also liberalized consolidations, particularly of neighboring fields but also of operations in general. The increasing depth and increasing cost of mining demanded consolidation of effort, finances, and fields. In 1852 the editor of the mining journal, Zeitschrift für Berg-, Hütten-, und Salinen Wesen suggested, "In order to compete with the large operations, the mine owners of smaller operations should consolidate their operations, or take the consequences."[8] He gave one of the clearest contemporary economic arguments for large scale operation:

> Only through the concentration of capital and labor into large operations can the general costs be reduced, the prices of mining products be lowered, and the mine owners still receive larger profits. 9

3. The Joint-stock Company

Between 1851 and 1865 the large joint-stock company and with it the large scale mining operation came to dominate the Ruhr coal industry. As Table 6.14 at the end of this section indicates, between 1854 and 1857 about 35 joint-stock companies were established in the Ruhr region. Nearly all of these were directly involved in coal mining activities. While the mining company, the Gewerkschaft, continued to serve the traditional mining interests and remained the typical organization for the small and middle sized operation, the joint-stock company with limited liability attracted the non-regional and especially foreign interests who were more familiar with this form of organization. Furthermore the Gewerkschaft was handicapped since the badly needed financial reform of the internal structure of this traditional mining organization was not accomplished until the General Mining Codification of 1865.

The unreformed Gewerkschaft was not adequate for the financial demands of the 1850's, but the joint-stock company attracting larger amounts of capital filled this need. The cost of acquiring fields, the cost of shaft construction as well as the cost of operation demanded increasing amounts of capital. The cost of acquiring fields went up dramatically during the 1850's. The ministerial order of 1853 had been intended to make the acquisition of mining property less costly by accepting the results of drilling as evidence for a claim. Still the price of fields had constantly been increasing. For instance, in 1854 Concordia paid for 15 fields, 250,000 thalers, but already in 1855, the Magdeburger Bergwerks-Aktiengesellschaft paid 120,000 thalers for only three fields and the A.G. Arenberg had paid 220,000 for four fields.[10]

Speculation during the boom period in the 1850's was a major cause for the high price of fields. In February 1855 three claims were bought for 60,000 thalers and sold in April of the same year for 120,000 thalers. Another example involved the A.G. Consolidation. The merchant A.W. Stoltenhoff sold seven claims on September 11, 1856, for 162,000 thalers to the merchant G.A. Stoltenhoff, who in turn sold them on the same day to Consolidation for 220,000 thalers.[11]

The cost of excavating shafts also increased. This expense frequently represented the most risky and costly part in a mining enterprise. The risk was high because the geological structure of the coal layers was still uncertain and in mining unforeseen costs arose, especially in coping with the water problem. For instance a water pumping steam engine might cost from 10,000 to 30,000 thalers. Consolidation paid 18,000 thalers, the A.G. Neu-Essen, 22,000 thalers.[12]

The length of time required to construct a shaft was normally several years. The newly formed joint-stock companies of the early and middle 1850's started their production only in the late 1850's or early 1860's. The entrance of the large firms during a time of severe depression accounted for the relatively rapid increase in production as well as the further rapid decrease in prices. It took the following joint-stock companies several years before they could actually market coal: Dahlbusch from 1851 to 1861;[13] Consolidation from 1848 to 1864; Arenberg from 1856 to 1861; Concordia from 1850 to 1854;[14] Magdeburg from 1856 to 1860; Neu-Essen from 1855 to 1859;[15] Harpen from 1856 to 1859. All of these had an initial capital of between 500,000 and 1,000,000 thalers.

During the 1850's, when the capital needs of the Ruhr mining industry could no longer be locally supplied, investors from large German cities and foreign countries participated.[16] For example the mine Consolidation had only seven shareholders, two from Essen and one each from the cities Cologne, Magdeburg, Königsberg, Hamburg and Braunschweig. The 24 shareholders of the A. G. Neu-Essen came from the cities Essen, Magdeburg, Hildesheim, Hannover, and Braunschweig, while in the case of the A. G. Arenberg the shareholders came from the following cities: 57 from Berlin, 16 from Aachen, 7 from Cologne, 5 from Hamburg, 5 from Düsseldorf.

During the 1850's British, Belgian, Dutch and French capital was attracted to the Ruhr region. The major intermediary for the foreign capital was the banking firm of Sal Oppenheim Inc. et Cie in Cologne, with its connections in the financial centers of Paris and Brussels. Foreign capital participated in many of the newly formed joint-stock companies. Even the names of some of these enterprises indicate foreign origin (all were among the largest mine operations by 1864): Neu-Schottland, Erin, Hibernia,[17] Holland, Shamrock,[18] Duetsch-Hollandisch A.G., Hütten-Aktiengesellschaft Leopold. The shareholders of the _Société Anglo-Belge des Charbonnages du Rhin_ were mostly from Brussels, some French interests, but also included the mayor of Birmingham, England. The Irishman W. Th. Mulvany not only participated in Shamrock, Hibernia, and Erin, but also settled in the Ruhr and "became truly one of us."[19]

The history of the two shaft mines Rheinelbe and Alma between 1846 and 1873 reflects the financial and economic complexity in the Ruhr region.[20] Rheinelbe consisted of 4 claims: (1) the Mühlheim Anfahrt claim of March 1845, by the farmer Anton Humann; (2) the Antonius claim of March 1856, by the estate trustee Franz Antonius; (3) the Gerhard claim of April 1846, by the merchant Gerhard Schmitz; (4) the Rudolph Heinrich claim of May 1846, by Mathias Waltenhoff from Mühlheim. In each case the finances were lacking to construct the required shaft for each claim. Consequently, the claims were consolidated on January 1848 under the name Rheinelbe. But due to the political disturbance of 1848 and financial problems no shaft was constructed. In September 1853 the four above mentioned parties sold three-fourths of their interest to the French entrepreneur Louis Henry Obert of Paris and to the Englishman John Brown of London. Both were to provide the necessary finances to build a shaft, but this did not happen. In March 1854, Obert and Brown sold their parts to the financially capable French firm Société des mines et fonderies du Rhin, Detillieux et Cie, forerunner of the Phönix A. G. für Bergbau and Hüttenbetrieb, an ironworks in Borbeck near Essen which had been established with French and Belgian capital.[21]

In the meantime the concession of fields had been made easier by the Order of November 1853 allowing the results of drillings. Already by 1854 there were 70 drilling operations in the Essen area.[22] The immediate result was a dramatic increase in mining activity in the region, as Table 6.15

indicates.

The new owner of Rheinelbe immediately took advantage of the order and received the concession for Rheinelbe without the construction of a shaft. In 1855 Charles Detillieux consolidated his Rheinelbe and the still unworked Alma with the recently reformed Phönix A. G.[23] In 1855 the construction of a shaft for Rheinelbe was begun, but because of water difficulties and financial problems, Detillieux decided in 1856 to sell both Rheinelbe and Alma to the *Gelsenkirchen Kohlengruben Gesellschaft*, which sold them back to him personally in 1857. A new firm was formed, the *Gelsenkirchen Bergwerksverein, Detilleux freres et Cie, Kommandit-Gesellschaft auf Aktien*. Detillieux again started the construction of a shaft. The construction was difficult, but finally in December 1861 operation of Rheinelbe began though the Alma had to wait until 1872.[24]

C. The Marketing and Consumption of Coal

During the period 1850-1865 coal mining turned to mass production. The large scale mine changed the structure of the industry and dominated its output. Mass production made coal itself a mass good. Sales and consumption of coal also underwent dramatic changes. Technical inventions and innovations in the coal-using industries turned coal into a means of production. The iron and steel industries became large and permanent consumers of coal when these industries

TABLE 6.14

35 NEWLY ESTABLISHED JOINT-STOCK COMPANIES: 1854-1857

1854, Mühlheimer Steinkohlen Bergbau Kommanditgesellschaft
Albert de Gruyter und Co.
Bergbaugesellschaft Neu-Essen
Bochumer Verein für Bergbau und Gusstahlfabrikation
Hörder Bergwerks-und Hütten Verein

1855, Gewerkschaft Hibernia
Gewerkschaft Erin
Ruhrort Mining Co.

1856, Harpener Gesellschaft für Kohlenbergbau
Dortmunder Bergbau und Hüttengesellschaft
Tremonia Bergbau-AG
Mark Bergbau-Gesellschaft
Berg-und Hüttengesellschaft
Arenbergische AG für Bergbau und Hüttenbetrieb
Broicher Bergwerks-AG
Aktiengesellschaft für Hüttenbetrieb und Bergbau
Kohlenbergbau-Gesellschaft Wilhelmine Victoria
Kohlenbergbau-Gesellschaft Caroline, Hüttengesellschaft Leopold
Bergbaugesellschaft Holland
Magdeburger Bergwerks-AG
Bergbau-Aktiengesellschaft Glückauf
Vollmond-AG
Gewerkschaft Ruhr und Rhein
Deutsch-Holländischer Bergwerksverein
Bergbau-Gesellschaft von Medio-Rhein
Rheinische Bergbau-und Hüttengesellschaft
Neu-Duisburg Bergbau-AG

1857, Bergbau-Ag Borussia
Apperbecker Aktienverein für Bergbau
Steinkohlen-Bergbau-AG
Venus und Potsdam Kohlenbergbau-gesellschaft
Hellweg Kohlenbergbau-Gesellschaft
Ruhrorter Bergwerks-AG
Niederländische Gesellschaft

Source: Däbritz, "Entstehung und Aufbau des rheinisch-westfälischen Industriegebiets," _Beiträge zur Geschichte der Technik und Industrie_, pp. 13-108.
Also see note 25.

TABLE 6.15

ACTIVITY IN MINING 1852-1860

Field Concessions	1852	1853	1854	1855	1856	1857	1858	1859	1860
(in existence)	929	982	1061	1129	1157	1252	1408	1470	1497
Search Warrants:									
Applied	447		486	431*	441*	411	105	29	9
Granted	358		406	368*	396*	226	43	13	6
Claims									
Applied	409		310	238*	294*	376	224	128	85
Granted	182		144	153*	197*	225	114	78	50
New Field concessions	138		46	68*	51*	90	185	202	83
Consolidations									
Number	8		N/A	N/A	6*	8	7	21	16
No. of Fields Consolidated	N/N		N/N	N/N	47*	33	29	100	59

*Only for the Bochum mining district (not Essen)

Source: <u>Zeitschrift für Berg-, Hütten-, Salinenwesen</u>, Vols. I-X.

shifted from the old charcoal furnaces to new coke furnaces. Also the increasing number of steam engines in all industries in the Ruhr region used coal. Tables 6.16 through 6.21 at the end of this section indicate the marketing and consumption pattern between 1850 and 1865.

The iron and steel industries were major consumers of coal. Their consumption of coal increased from 6.8 per cent of total Ruhr output in 1851 to 17.0 per cent in 1856. The depression affected the iron and metallurgical industries most severely; consequently coke and pig iron production suffered. In 1852 only two of the 12 blast furnaces used coke and produced only 9.6 per cent of the Ruhr pig iron production. In 1865 all 51 blast furnaces used coke in the production of pig iron. By 1865 the Ruhr region produced more than one third of all Prussian pig iron (Tables 6.16 and 6.17).

From 1850 to 1865 the Ruhr coal had attracted large and permanent consumers not only inside the Ruhr region but outside as well. With the improvement in transportation through the new steamships and railroads the market outside the Ruhr expanded considerably. During the 1850's the railroad became the major means of transporting coal outside of the Ruhr region. In 1851 only 24.9 per cent of coal was marketed by the newly built railways, but in 1861 it was already nearly 60 per cent. The Cologne-Minden and the Berg-Mark railways were the major coal transporters (Tables 6.18 through 6.21), and the largest shaft mines, particularly in the Essen region, had their own connections with these lines. In addition the railroads themselves became constant consumers of coal, using annually around 3 per cent of the coal production.

The other major means of transportation was water transport. The Ruhr River transport decreased from nearly 30 per cent in 1851 to about 15 per cent in 1865. On the other hand, the Rhine remained as the major non-regional artery. The introduction of steamships and large iron barges became important for the export of Ruhr coal (Table 6.22). The export of coal from the two ports, Duisburg and Ruhrort although it increased, was, nevertheless, a decreasing share of the total Ruhr production. The railroad successfully competed (Table 6.23).

The foreign coal market also expanded (Table 6.24). Holland was a major market but also transhipment of coal through Dutch ports probably increased. Coal shipment to the Netherlands had increased from 1851 to 1860 from 170,000 tons to 430,000 tons respectively. Coal was also transported upstream as far as Switzerland. Shipment on the Rhine had received encouragement by the abolition of transit duties on the river in 1854.[26]

In general the reform of the Direktionsprinzip and the economic changes described in this chapter laid the foundation for the modern Ruhr industrial region. During the

1850's the era of direct and detailed state control came to an end. There remained only the writing of the formal obituary in the form of the General Mining Codification of 1865.

TABLE 6.16

COKE PRODUCTION AND USE IN BLAST FURNACES
1850-65

Year	Number of blast furnaces using coke	charcoal	mixed	total	Coke Production (1,000 tons)
1850	-	-	-	15	73
1851	-	-	-	16	79
1852	2	8	2	12	114
1853	4	8	2	14	112
1854	19	4	3	26	188
1855	21	6	2	29	224
1856	27	7	2	36	228
1857	25	6	2	33	222
1858	22	8	1	31	216
1859	22	7	2	31	182
1860	19	6	2	27	198
1861	-		0	33	
1862	-		-	39	
1863	-		-	45	
1864	-		-	45	
1865	-		-	51	

Source: *Zeitschrift für Berg-,Hütten-,Salinenwesen*, Vol. 1-13

TABLE 6.17

COKE AND PIG IRON PRODUCTION: 1850-1865

Years	Ruhr Percentage Pig Iron Produced with coke	charcoal	mixed	Pig Iron Prod. 1,000 tons	Ruhr Percentage Pig Iron of Prussia
1850	9.6	63.0	27.0	11	8.5
1851	40.0	34.0	25.0	12	8.3
1852	60.0	20.0	19.0	22	13.4
1853	68.0	15.0	17.0	31	14.7
1854	85.0	10.0	4.0	54	20.5
1855	90.0	7.0	2.8	86	28.5
1856	95.0	4.0	1.0	122	33.6
1857	95.0	3.5	1.5	140	35.2
1858	94.0	5.1	Below	130	31.5
1859	94.0	4.1	one	121	30.6
1860	95.0	3.8	per	136	34.5
1861	97.0	2.9	cent	151	33.6
1862	98.0	1.9	-	191	36.4
1863	98.0	1.4	-	231	36.3
1864	99.0	1.0	-	252	35.7
1865	99.0	1.0	-	272	35.8

Source: <u>Zeitschrift für Berg-, Hütten-, Salinenwesen,</u> Vols. I-XIII.

TABLE 6.18

COAL CONSUMED BY IRON AND STEEL WORKS, 1851-1856
(PER CENT OF TOTAL COAL PRODUCTION)

Consumption by iron and steel works in Ruhr

1851	1852	1853	1854	1855	1856
6.8	8.6	10.2	13.1	15.8	17.0

Source: *Zeitschrift für Berg-, Hütten-, Salinenwesen*, Vols. I, II, III.

TABLE 6.19

SALE OF COAL USING THE RUHR, RAILROAD AND OTHER MEANS: 1851-1861 (PERCENT)

Year	1851	1852	1853	1854	1855	1856	1857	1858	1859	1860	1861
By Ruhr	29.6	29.6	29.1	27.1	23.2	20.1	15.7	14.6	15.0	16.7	14.7
By Railroad	24.9	28.3	30.0	33.0	36.7	40.9	45.3	48.3	50.3	55.1	59.5
Other	38.5	35.3	34.4	33.4	34.1	32.5	32.3	30.2	26.8	20.7	18.2
Consumption at mines	7.0	6.8	6.7	6.5	6.0	6.5	6.7	6.9	7.9	7.5	7.6

Source: *Zeitschrift für Berg-, Hütten, -Salinenwesen*, Vols. I-VIII.

TABLE 6.20

COAL TRANSPORT ON THE COLOGNE-MINDEN AND BERG-MARK RAILWAYS: 1854-1861

Year	tons (1,000)
1854	990
1855	1,240
1856	1,385
1857	1,795
1858	2,265
1859	2,410
1860	2,820
1861	3,005

Source: Same as Table 6.19.

TABLE 6.21

TRANSPORT OF COAL ON THE COLOGNE-MINDEN AND BERG-MARK RAILWAYS AND THEIR END POINTS: 1855-57

	1855	1856	1857
To Cologne	785,000	840,000	1,005,000
To Minden	165,000	180,000	150,000
To Elberfeld	170,000	180,000	430,000
To Soest	120,000	185,000	210,000

Source: <u>Zeitschrift für Berg-, Hütten-, Salinenwesen</u>, Vols. II-IV.

TABLE 6.22

STEAMSHIPS ON THE LOWER RHINE: 1850'S

Steamship Company	Steamships			Barges	
	Number	Horsepower	Capacity (tons)	Number	Capacity (tons)
Netherlands:					
Rotterdam Co.	5	780	468	5	1,160
Dordrechter Co.	2	500	108	-	-
Prussia:					
Cologne Co.	4	750	172	32	10,860
Düsseldorf Co.	4	600	183	12	3,850
Mühlheim Co.	5	1,400	302	-	-
Ruhrort Co.	6	1,650	264	-	-
Franz Haniel	4	1,070	237	11	4,730
Mathias Stinnes	2	550	201	-	-

Source: Hans Spethmann, Ruhrgebiet, Vol. II, p. 313.

TABLE 6.23

EXPORT OF COAL FROM THE PORTS OF DUISBURG AND RUHRORT TO THE RHINE, 1850-65

Year		Total Production
1850	802,900	1,666,000
1855	1,128,200	3,310,000
1860	1,319,000	4,360,000
1865	1,867,300	9,270,000

Source: Zeitschrift für Berg-,Hütten-,Salinenwesen, Vol. VIII.

TABLE 6.24

WATER TRANSPORT OF COAL FROM RUHRORT/DUISBURG PORTS ON THE RHINE: 1851-60

(1,000 tons)	1851	1852	1853	1854	1855	1856	1857	1858	1859	1860
To Dutch border	100	104	110	120	130	420	370	110	90	90
Into the Netherlands	170	205	310	306	340	280	200	280	310	430
Upstream	390	490	510	610	650	380	190	730	650	800
Total	660	899	830	1066	1120	1080	760*	1120	1050	1320

*Low water on Rhine

Source: Zeitschrift für Berg-,Hütten,-Salinenwesen, Vols. I-VII.

FOOTNOTES - CHAPTER VI

1. Heinrich Achenbach, "Über das Bergregal und Berghoheitsrecht in Preussen," <u>Z.f.B.H.S.-W.</u>, XIII, 76.

2. "Bergwerksbetriebe in dem preussischen Staate," <u>Z.f.B.H.S.-W.</u>, I (1852), 65.

3. <u>Niederrheinisches Industriegebiet</u>, X, 33-34.

4. Krampe, p. 70.

5. Z.f.B.H.S.-W., I (1854), 67.

6. <u>Ibid</u>., p. 66.

7. <u>Z.f.B.H.S.-W.</u>, II (1855), 197-198.

8. <u>Z.f.B.H.S.-W.</u>, I, 66.

9. <u>Ibid</u>, I, 66.

10. Däbritz, pp. 41-49.

11. <u>Ibid</u>., p. 41.

12. <u>Ibid</u>., p. 42.

13. Dahlbusch was already the 24th largest mine, when it started production in 1862, see <u>Z.f.B.H.S.-W.</u>, XI (1863), 115-117.

14. <u>Ibid</u>., Concordia was 10th largest mine in 1862.

15. <u>Ibid</u>., 18th largest mine in 1862.

16. Däbritz, "Entstehung...," p. 46.

17. Hibernia was the largest operation in the Mark region in 1860, <u>Z.f.B.H.S.-W.</u>, IX, 65, and second largest in the Ruhr region.

18. Shamrock, 9th largest operation in Ruhr region in 1861, X, 70-73.

19. Däbritz, p. 47.

20. G. Gebhardt, Ruhr berg bau, pp. 120-190.

21. See Cameron, pp. 374-375, Gebhardt ibid.

22. Hans Spethmann, Gron westschraft an der Ruhr, 1925, p. 122.

23. See Cameron, pp. 374-375, for French capital involvement in Rheinelbe.

24. Rheinelbe was the 7th largest operation in 1864, Z.f.B.H.S.-W., XII, 116.

25. Däbritz, pp. 13-108; As even many of the names indicate, foreign capital was involved.

26. Cameron, France, p. 18.

CHAPTER VII

THE DIRECTIONSPRINZIP IN HISTORICAL PERSPECTIVE

A. The Codification of 1865: The End of the Direktionsprinzip

The General Prussian Mining Codification of June 24, 1865, unlike previous laws and codifications, was simply worded and clearly stated.[1] It basically contained nothing that the reform laws had not already instituted or intended. The one major exception, discussed later in this section, was the reform of the share system in the traditional mining company.

The codification dealt essentially with four major topics: (1) the aquisition of mining property; (2) the mining operation; (3) the mining company and its internal structure; (4) the police functions of the mining administration. The rest of the codification dealt with the property owner, the labor union, the hierarchy of the mining administration, abolition of the older laws, and transitional regulations.

Part II of the codification focused on the acquisition of mining property: the procedures of prospecting for minerals, claiming a field, securing a concession, and consolidation of concessions.[2] Prospecting for minerals was free for everyone, except where it would be detrimental to the general public. Permission was required from the property owner who had to be compensated but could not prohibit prospecting on his property. The claiming of a field involved the formal request to the mining office for the concession of a specified field.[3] This request was granted after the mineral had been proven to exist. The coal discovered had no longer to be proven worthy of mining and literally uncovered to eyesight. The result of drillings or other evidence, such as neighboring fields, was sufficient to prove the existence of coal.

A concession was normally granted after the claim had been registered and the field had been precisely determined.[4] The concession was granted by the mining office. Consolidation of fields or mining operations was also under the jurisdiction of the mining office, but no conditions or qualifications were attached other than following the proper legal procedure.

The second major topic concerned the mining property: its technical plan of operation, management, and employment. After a concession had been formally granted the mining property became immobile property. Once an operation started, the mining office could halt the operation only in the interests of the general public, and the mining office had the power under extreme circumstance to confiscate the mining property.[5] A mine could only operate according to a technical plan.[6] This plan had to be checked and approved by the mining office. However, the examination of the plan was strictly limited to police and safety standards. But a mine had to operate according to such a plan or it could be closed.

The strongest control maintained by the mining administration under this codification concerned the qualifications of the technical personnel employed by the private company.[7] The technical personnel, the operation's manager and technical supervisors, were obliged to prove their qualifications by taking an examination. Only after they had proven their qualification were they allowed to assume their positions. The mining office had the right to remove unqualified persons and if necessary to halt the operation until a capable person was employed.

There was freedom of contract between mine workers and mine owners, but the contract was defined and supervised by the mining office.[8] Both sides had to adhere to certain formal conditions such as giving notice, and the employer was required to provide a written evaluation of the released miner. There were further legal safeguards for the worker.

The third major topic dealt with the Gewerkschaft and its internal structure: legal representation, decision making, and the share system.[9]

Two or more participants could form a company and a three fourths majority of the shares could determine the legal form of the company. Each company had to have a representative or a board (which could consist of non-shareholders) who were elected by an absolute majority of the shares at a shareholders' meeting. The representative or the board had full legal powers to act on all matters in behalf of the company. Import decisions were made by the shareholders at an official meeting. A three fourths majority of shares could make all the necessary decisions concerning leasing, exchanging of mining property, bankruptcy, and sales of the mining operation. Finally the share system of the Gewerkschaft was reformed.

Since the 1830's the reform of the legal nature of the mining company and its financial structure had been advocated, but all attempts had failed. Even the 1851 laws did not reform the nature of the share system; nevertheless, the issue of "the mobilization of the share (Kuxe)" was kept alive during the 1850's. A new attempt was made in 1861 with a mining law proposal specifically dealing with the share system. This revision proposed the transferable share and the recognition of the mining company as a legal personality. Indirectly the 1851 law had already done the latter. Since the representatives of a company were held legally responsible for the entire company, they could sue and be sued in behalf of the company.

The 1861 revision failed due to differences of opinion not only within the mining administration but also from strong opposition by the traditional mining interests within the Ruhr region. Furthermore the revision came at a difficult time, the height of the depression. To many mine owners and shareholders the traditional share system of the mining company appeared to provide financial security which the joint-stock company could not offer. In fact, the shareholders of the Gewerkschaft used the conditions of the depression as an argument against the transferable share proposed in the 1861 revisions. They pointed to the financial problems of the newly formed joint-stock companies.

The shareholders of these new organization had not been required to pay the entire amount of their shares at the time of purchase. The payments were usually 5 or 10 per cent of par value.[10] But as the depression advanced the first payment was raised to 20 and even 40 per cent.[11] This was in many cases beyond the financial ability of the individual shareholder. Consequently, some joint-stock companies liquidated, foreign capital withdrew, and entire operations were halted. Furthermore, many shareholders had taken loans, using their shares as securities.

Klostermann, a mining law expert defending the revision,[12] argued that the debt on the shares as of June 1, 1860, was 11.8 million thalers or only 57 per cent of the average yearly gross income by the mining companies in Prussia. In the Ruhr region it was 4.5 million thalers, but one joint-stock company alone had 1.5 million thalers yet to pay to the original owner. This amount, however, was entered as a debt against the existing shares. According to Klostermann's estimates such unpaid debts by mining companies constituted somewhat more than one half of the debt listed for 1860.

"The acquisition of property, rather than the economic crisis, accounted for most of the present debts [1860]."[13] Furthermore Klostermann argued against those who claimed that most of the capital had come from the personal wealth of the shareholders in the form of _Zubusse_ and loans to the company. He claimed such monies amounted to only a few million thalers in 1860, a small percentage of the total capital required by the mining industry.

Despite the failure of the 1861 proposal, the work on a general mining codification had begun, and Part IV of this codification finally reformed the financial structure of the mining company.[14] The mining company was recognized as a legal personality and it alone could assume debts and was liable for its assets. Individual shareholders could not claim part of the company. The number of shares was set at 100 but according to the by-laws the company's shareholders could increase the number of shares to 1,000. The share was not divisible and had the character of mobile property.[15]

One traditional characteristic of the old mining company was maintained in that the shareholder participated in the profits but also in the losses of the company. The shareholder still was required to pay _Zubusse_. Otherwise, each shareholder could transfer his shares without the permission of the rest of the shareholders. This transfer of certificates merely had to be registered in the company's books and the new owner was responsible for the obligation of the shares.

The fourth major topic dealt with the police function of the mining administration.[16] The mining office had police jurisdiction over: (1) the safety of the mining operation, (2) the protection of the life and health of the workers; (3) the safeguarding of the surface in the interest of personal safety and public transportation, (4) protection of the general public from harm by the mining industry. The rest of this section of the mining codification dealt with specific rights and duties of the mining administration in regard to safety measures protecting the general public.[17]

Finally, it should be clear, from the above, the General Prussian Mining Codification of 1865 did not introduce a complete *laissez-faire* policy. The state maintained important and meaningful controls over the industry.

B. The *Direktionsprinzip* in Perspective

The 1766 Cleve-Mark Mining Codification had introduced the *Direktionsprinzip*; the 1865 General Prussian Mining Codification ended the strong state controls over the Ruhr mining industry, reducing these controls to police and inspection power. During the hundred years, the intervention of the Prussian state in the economic life of the mining industry did result in its development. The mining law had provided a framework for the growth of the industry as well as the region itself. The state intervention had also led to institutional and political changes which eventually established a market economy in the Ruhr mining industry.

The state's policy and its implementation had involved the state in a partnership with private enterprise. Of course, during the 100 years the state's dominant role in this relationship did not always represent the best interests of the Ruhr mining industry nor of the region itself. At times, the state acted as an obstacle to continued growth. In terms of the state policy, its legal implementation under the *Direktionsprinzip*, and the economic growth of the Ruhr mining and related industries, the 100 years can be divided into three fairly distinct periods: (1) 1766-1815; (2) 1815-1850; (3) 1851-1865.

During the years from 1766 to 1815 the mining industry experienced the implementation of the *Direktionsprinzip*. The state pursued an active and progressive policy, assuming a dynamic and leading role in the industry and the region. The mining administration brought order, organization, and method to private enterprise and the industry. Under this administration the state supplied the necessary qualified personnel and technical knowledge to operate the industry.

In the absence of mature private enterprise, the state assumed the role and function of private enterprise. Furthermore, the *Direktionsprinzip* served as the basis of an overall economic policy promoting the growth of the Ruhr mining region itself. The state provided the infrastructure, building roads and particularly improving the navigation of the rivers. In general the state initiated economic policies and programs fostering the growth of the Ruhr mining industry. The state not only achieved the economic integration but also the political unification of the entire Ruhr region.

During the period 1815-1850 the state no longer exercised its vital role in the development of the industry or the region. Political, economic and institutional conservatism tended to dominate. The *Direktionsprinzip* was maintained, although it was clearly losing its effectiveness. This system of state control and its administration calcified and actually became a hindrance to the growth of the industry. Private enterprise rather than the state assumed the active role. During this period such growth as took place in the industry was not because of the *Direktionsprinzip*, but rather in spite of it. Furthermore, it was private initiative and enterprise which introduced and developed the new transportation facilities, the railways. Not until the end of the period did the state become involved in transportation.

In general this was a period of tension between the state and private enterprise. The *Direktionsprinzip* had outlived its usefulness and particularly needed reform after 1830. Reform was proposed but despite the mounting pressures the state was unwilling to change this archaic system. The state's policy had gradually become itself an obstacle which needed to be removed for the continuous and rapid economic growth of the industry.

During the period 1851-1865, the Ruhr mining industry experienced the reform of the *Direktionsprinzip* and its most rapid economic growth. Political and economic pressures had forced the state to withdraw its tutelage and direction of the industry. Now the market rather than the state regulated the growth of the industry. The state, however, did maintain a minimum of controls to protect the public and to assure safety and a standard of technical efficiency in the industry. It is interesting that under Freiherr vom Stein, considered a liberal reformer, the *Direktionsprinzip* was fully implemented in the Ruhr region, and that another liberal reformer, August von der Heydt, was chiefly responsible for abolishing it.

FOOTNOTES - CHAPTER VII

1. "Allgemeines Berggesetz für die Preussischen Staaten, 24. Juni 1865," Z.f.B.H.S.-W., XIII (1865), 29-65.

2. Ibid., SS 3-11.

3. Ibid., SS 12-21.

4. Ibid., SS 22-38.

5. Ibid., SS 50-64.

6. Ibid., SS 67.

7. Ibid., SS 73-79.

8. Ibid., SS 80-93.

9. Ibid., SS 94-127.

10. "Statuten der Anonymen Gesellschaft Phönix," Z.f.B.H.S.-W., II (1855), 110-116

11. Däbritz, p. 149.

12. Klostermann, pp. 323-325.

13. Ibid., p. 325.

14. General Prussian Mining Codification of 1865, Z.f.B.H.S.-W., XIII, part IV, articles 97 to 104.

15. Ibid., Article 101.

16. Ibid., SS 196-209, SS 153-155.

17. Ibid., SS 196.

BIBLIOGRAPHY

Achenbach, Heinrich, Das französische Bergrecht, Bonn: 1869.

______________, "Geschichte der Cleve-Märkischen Berggesetzgebung und Bergverwaltung bis zum Jahre 1815," Zeitschrift für Berg-,Hütten-Salinenwesen, J.g. 1869.

______________, Das gemeine deutsche Bergrecht in Verbindung mit dem preussischen Bergrechte, Bonn: 1871.

Adlemann, Gerhard, Die Soziale Betriebsverfassung des Ruhrbergbaus vom Anfang 19. Jahrhunderts bis zum ersten Weltkriege, Bonn: 1962.

Administration Law: "Gesetz über die Aufsicht der Bergbehörden über den Bergbau und das Verhältnis der Berg-und Hüttenbetriebe, 21. Mai 1860." Zeitschrift für Berg-Hütten-Salinenwesen (1860).

Adolphe, A. Ruhrkohlenbergbau, Transportation und Eisenbahntarifpolitik, Berlin: 1827.

"Aktienvereine für Bergwerks-und Hüttenbetriebe im Preussischen Staate," Z.f.B.H.S.-W. (1854).

Arlt, H. Ein Jahrhundert preussischer Bergverwaltung in den Rheinlanden, Bonn: 1921.

Arndt, A. Bergbau und Bergbaupolitik, Leipzig, 1894.

______________, Zur Geschichte und Theorie des Bergregals und der Bergbaufreiheit, Freiburg, 1916.

Augustin, H.F.L., Das Preussische Handels-Ministerium in seinem Verhaeltnisse zu den Privat-Eisenbahn Gesellschaften, Potsdam, 1859.

Arzberge, M., Eisenbahnen als Staats und als Gesellschaftsunternehmungen, Hamburg: 1842.

Barhold, Karl, Wahrnehmungen bei der Entwicklung der Transportmittel, Berlin: 1886.

Banfield, T.C., The Industry of the Rhine, London: 1846.

Barleben, Ilse, Mühlheim a.d. Ruhr, herausgegeben von der Satdt Mühlheim, 1959.

Baumann, Hans, Deutsches Verkehrbuch, Berlin: 1931.

Bechtel, Heinrich, Wirtschaftsgeschichte Deutschlands, im 19. und 20. Jahrhundert, Müchen: 1956.

Beckerath, E.V., Die Preussische Klassensteuer und die Geschichte ihrer Reform bis 1851, München; 1912.

Beiträge zur Geschichte der Technik und Industrie (Jahrbuch des Vereines deutscher Ingeneure) Herausgegeben von Conrad Matschoss, Berlin.

Bergengrün, A., David Hansemann, Berlin: 1906.

______________, Staatsminister August von der Heydt, Berlin: 1908.

Berger, L., Der alte Harkort, Leipzig, 1890.

Bergleute: Reglement für Bergleute, 30 April 1781, Mining Library Essen: 8394a 2, Photocopy.

Bergrechtliche Blätter, Beilage zur Östereichischen Zeitschrift für Berg-& Hüttenwesen, Wien: Vierteljahrschrift herausgegeben von Ludwig Haberer, 1908-1914.

Bloch, Adolf, Die Berggesetzgebung der Französischen Revolution, Berlin: Akademie Verlag, 1954.

Boldt, Gerhardt, Staat und Bergbau, München: 1950.

Botzenhart, Eric, Freiherr vom Stein, Münster Westf. Aschendorff: 1952.

Brassert, Hermann, Allgemeines Berggesetz für die Preussischen Staaten vom 24. Juni 1865, mit Einführungsgesetz und Kommentar, Bonn: 1888.

______________, Bergordnungen der Preussischen Lande, Köln: 1858.

______________, Das Bergrecht des Allgemeinen Preussischen Landrechts in seinen Materialien, Bonn: 1861.

______________, "Die Bergrechtsreform in Preussen." Zeitschr. für Bergrecht, Jg. 1862.

Brepohl, Friederich Wilhelm, _Industrievolk, im Wander der agraren zur indusriellen Daseinsform dargestellt am Ruhrgebiet_, Tübingen, 1951.

Cameron, Rondo E., _France and the Economic Development of Europe, 1800-1914_, Princeton: 1961.

Carnall, R.v., _Die Bergwerke in Preussen und deren Besteuerung_, Berlin, 1850.

_____________, _Die Bergwerksverhältnisse in dem preussischen Staate_, Berlin, 1856.

_____________, "Das Denkmal des Ministers Grafen von der Reden bei Königshütte," _Z.f.B.H.S.-W._, (1854).

Chevalier, Michael, trans. Lindner, Friederich Ludwig, _Die Eisenbahnen im Vergleich mit den Wasserstrassen_, Stuttgart & Tübingen, 1838.

Clapham, John H., _The Economic Development of France and Germany 1815-1914_, Cambridge: 1961.

Clauss, Herbert, "Rechtsbestimmungen für die bergmännische Arbeit im 12, 13, 14 Jahrhundert", _Freier Berg und Vermessenes Erbe_, Berlin: Akademie Verlag, 1957.

Cohn, Gustav, _Zur Geschichte und Politik des Verkehrswesens_, Stuttgart: 1900.

Conze, W., "Die Wirkungen der liberalen Agrarreform auf die Volksordnung in Mitteleuropa im 19. Jahrhundert," _Vierteljahrsschrift für Sozial-und Wirtschaftsgeschichte_, 1949.

Däbritz, Walter, "Das staatliche Direktionsprinzip," _Deutsche Bergwerks-Zeitung_, 25. 8. 1924, Jubiläumsausgabe Nr. 2.

_____________, "Entstehung und Aufbau des rheinisch-westfälischen Industriebezirks", _Beiträge zur Geschichte der Technik und Industrie_, Berlin: 1925.

_____________, "Finanzprobleme aus der Entstehungszeit des rheinisch-westfälischen Industriereviers," _Glückauf_, Jg. 1922.

_____________, _50 Jahre Metallgeschaft 1881-1931_, Frankfurt: 1931.

Denkschrift des Komitees für Beförderung des Absatzes der westfälischen Steinkohlen nach dem Osten. Essen 1858, Bergbaubücherei Essen Nr. Fi 16, Fotokopic.

Dieterici, K., *Der Volkswohlstand im Preussischen Staate*, Berlin: 1846.

___________, *Handbuch der Statistik des preussischen Staates*, Berlin: 1858.

___________, *Statistische Uebersicht der wichtigsten Gegenstande des Verkehrs und Verbrauchs im preussischen Staat und im Zollverein, 1838-1857*, Berlin: 1858.

Ebert, Kurt, *Agricola und das Bergrecht*, Freiberg: 1953.

Eckert, C., *Rheinsschiffahrt im 19. Jahrhundert*, Leipzig: 1900.

Eichholtz, Dietrich, *Junker und Bourgoisie vor 1848 in der Preussischen Eisenbahngeschichte*, Berlin: 1962.

Ermisch, Hubert, *Urkundenbuch der Stadt Freiberg in Sachsen*, Leipzig: 1883.

Ernst, Carl, "Uebersicht der Bergsetzgebung in Italien", *Bergrechtliche Blätter*, v. 4, 1909.

Fechner, Hermann, Geschichte des Schlesischen Berg- und Hüttenwesens in der Zeit Friedrichs des Grossen, Friedrich Wilhelms II. und Friedrich Wilhelms III. 1741-1806," *Zeitschr. für Berg-, Hütten- und Salinenwesen*, Jg. 1900, 1901, 1902.

Felderhoff--von Vincke,"Geschichtliche Darstellung und Entwicklung der Ruhrschiffbarmachung, Jan. 1833." (Staatsarchiv Münster: OBA = 37/14).

Ferber, C. *Neue Beiträge zur Kenntnis der gewerblichen und kommerziellen Züstande der preussischen Monarchie*, Berlin: 1832.

Fischer, Wolfram, *Die Bedeutung der preussischen Bergrechtsreform (1851-   ) für den industriellen Ausbau des Ruhrgebietes*, Dortmund: 1961. "Die Stellung der Preussischen Bergrechtsreform von 1851-1865 in der Wirschafts-und Sozialverfassung des 19. Jahrhunderts", *Zeitschrift fuer die Gesamte Staatswissenschraft*, Tübingen: J.C.B. Mohr, 1961.

Felsch, G., Die Wirtschaftspolitik des Preussischen Staates beider Gründung der Oberschlesischen Kohlen-und Eisenindustrie (1741-1871), Berlin: 1919.

Fleck, G., "Die ersten Eisenbahnen von Berlin nach den Westen der Monarchie," Archiv für Eisenbahnwesen, 1895.

______________, "Studien zur Geschichte des preussischen Eisenbahnwesens," Archiv für Eisenbahnwesen (19., 20. u. 21. Jahrg.) Berlin: 1896, 1897 und 1898.

General Codification of 1865, "Allgemeines Preussisches Berggesetz, 24. Juni 1865," Z.f.B.H.S.-W. (1865).

Gergius Agricola, 1494-1555, Berlin: Deutsche Akademie der Wissenschaft, 1955.

Gesetz-Sammlung für die Königlichen Preussischen Staaten, Berlin.

Gleim, "Zum 3 November 1838", Archiv fuer Eisenbahnwesen, 1888.

Gmelin, Johann, Beiträge zur Geschichte der teutschen Bergbaues, Halle: 1783.

Goltz, J. von der, Auswirkungen der Stein-Hardenbergischen Agrarreform im Laufe des 19. Jahrhundert, Berlin: 1936.

Gothein, E., Grundriss der Sozialökonomik, VI Abteilung: Industrie, Bauwesen, Bergwesen, Tübingen: 1923.

______________, Die Schiffahrt der deutschen Ströme. Untersuchungen über deren Abgabenwesen, Regulierungskosten und Verkehrsverhältnisse. Schriften des Vereins für Socialpolitik, Bd. 2. Leipzig: 1903.

Hansemann, D., Die Eisenbahnen und deren Aktionäre in ihrem Verhältniss zum Staat, Leipzig: 1837.

______________, Ueber die Ausführungen des Preussischen Eisenbahnsystem, Berlin: 1848.

Hartwich, Bemerkungen über, Transportmittel und Wege so wie über Gestaltung und Verwaltung des Eisenbahnwesen, Berlin: 1878.

Haussherr, H., Wirtschaftsgeschichte der Neuzeit, third edition 1960.

Helfferich, Karl, Germany's Economic Progress and National Wealth 1888-1913, New York: Germanistic Society of America, 1914.

Henderson, William, The State and the Industrial Revolution in Prussia 1740-1870, Liverpool: 1958; "The genesis of the industrial revolution in France and Germany in the 18th century", Kyklos, 1956.

Henz, L., Der Ruhrstrom und seine Schiffahrtsverhältnisse, Essen: 1840.

Herrmann, Walter, Bergbau und Kultur, Berlin: 1953.

Hocker, N., Die Grossindustrie Rheinlands und Westfalens, Leipzig: 1867.

Hoffmann, W.G., and Müller, J.H., Das Deutsche Volkseinkommen, 1851-1957, Tubingen: 1959.

Hue, O., Die Bergarbeiter, Stuttgart: 1910/13.

Hundert Jahre deutscher Eisenbahnen, Jubiläumsschrift zum hundert Jährigem Bestehen der deutschen Eisenbahnen, Leipzig: 1938.

Huyssen, "Das Rechtsverhältnis der in denselben Feldern beliehenen Eisenstein-und Steinkohlengewerkschaften," Z.f.B.H.S.-W. (1954).

Inama-Sternegg, Karl T., Deutsche Wirtschaftsgeschichte des 10-12 Jahrhunderts, Leipzig: 1891.

Jahresbericht der Handelskammer für Essen, Werden und Kettwig, pro 1858, pro 1859, pro 1860, pro 1861, Essen 1858-1860: Bergbaubücherei Essen.

Jahresbericht der Handelskammer Köln für 1857, 1858, 1859, 1860 Köln 1858-1861.

Jahresbericht der Handelskammer für den Kreis Bochum, Bochum 1859, Bochum 1860, Witten 1861: Bergbaubücherei Essen.

Jahrbuch für die amtliche Statistik des preussischen Staates, 1863-83.

Jahrbuch für Volkswirtschaft und Statistik, 1852-63.

Johannsen, Otto, Geschichte des Eisens, Düsseldorf: 1925.

Joint-stock Company Law: "Gesetz über die Aktiengesellschaften, 9, Nov. 1843," *Z.f.B.H.S.-W.*, (1855).

Jüngst, E., *Festschrift zur Feier des fünfzigjährigen Bestchens des Vereins für die bergbaulichen Interessen im Oberbergamtsbezirk Dortmund in Essen 1858-1908*, Essen: 1908.

Kech, E., *Geschichte der deutschen Eisenbahnpolitik*, Leipzig: 1911.

Karsten, C.J.B., *Grundriss der Deutschen Bergrechtslehre mit Rücksicht auf die französische Berggesetzgebung*, Berlin: 1828.

__________, *Über den Ursprung des Berg-Regals in Deutschland*, Berlin: 1844.

Kersten, C., *Die revidierte Bergordnung für das Herzogtum Cleve, Fürstentum Moers und für die Grafschaft Mark vom 29. 4. 1766, nebst den dieselbe ergänzenden neueren Bestimmungen*, Dortmund: 1856.

Kern, V., *Die wirtschaftliche Entwicklung der Ruhrschiffahrt*, Diss. Köln: 1921 (Bergbaubücherei Essen).

Klosterman, R., "Bemerkungen über die Mobilisierung der Kuxe," *Z.f.B.H.S.-W.*

"Die Knappschaftsvereine im Preussischen Staat," *Z.f.B.H.S.-W.*, (1855).

Projekt der Konsolidation der 20 swischen der Ruhr und dem Muttenbach in der westl. Fortsetzung der Borbecker Mulde belegenen Zechen und der auf dem rechten Ruhrufer lagernden, der Borbecker Mulde angehörigen Zechen. Bergbaubücherei Nr. Fi 143.

Krampe, Hans D., *Der Staatseinfluss auf den Ruhrkohlenbergbau in der Zeit von 1800 bis 1865*, Köln: 1959.

Krenkel, Paul, "Zu der Urkunde des Markgrafen Otto vom 2 Aug. 1185", *Bergbau und Bergleute*, Berlin: Akademie Verlag, 1955.

Kube, Siegfried, "Der Bergbau in der Werdenden markgräflichen meissener Landesherrschaft", *Freier Berg und Vermessenes Erbe*, Berlin: Akademie Verlag, 1957.

Kumpmann, K., *Die Entstehung der Rheinischen Eisenbahn-Gesellschaft 1830 bis 1844. Ein erster Beitrag zur Geschichte der Rheinischen Eisenbahn.* Essen: 1910.

Zeche Langenbrahm in ihrer Entstehung und Entwicklung 1772-1922, (Berbaubücherei Essen), 1922.

Lehmann, M., Freiherr vom Stein, Leipzig: 1902-1905.

Loescher, Hermann, "Der Landesherrliche Schiedspruch fom 4. Sept. 1469", Bergbau und Bergleute, Berlin: Adademie Verlag, 1955.

Lottner, Die Fahrkunst auf der Steinkohlengrube "Gewalt", Z.f.B.H.S.-W., (1854).

Lotz, W., Verkehrsentwicklung in Deutschland 1800-1900, Leipzig und Berlin: 1920.

Luetge, Friederich, Deutsche Sozial und Wirtschaftsgeschichte, 2d ed. Berlin: 1952.

Matschoss, Conrad, Die Entwicklung der Dampfmaschine, Berlin: 1908.

____________, Ein Jahrhundert deutscher Maschinenbau. Von der mechanischen Werkstätte bis zur Deutschen Maschinenfabrik, 1819-1919, Berlin: 1919.

____________, Preussens Gewerbeförderung und ihre grossen Männer, Berlin: 1921.

Mayer, Arthur von, Geschichte und Geographie der deutschen Eisenbahnen von ihrer Entstehung bis auf die Gegenwart, 2 vol., Berlin: 1891.

Martini, "Die Einführung des Einpfennigtarifs für die Beförderung oberschlesischer und westfäfischer Kohlen nach Berlin," Archiv für Eisenbahnwesen, (1890).

Mayer, Theodor, Deutsche Wirtschaftsgeschichte der Neuzeit, Leipzig: 1928.

Meuss, P., Die Entwicklung des Steinkohlenbergbaus in Westfalen besonders in technischer Beziehung bis zum Jahre 1865, Berlin: 1930. (Sonderabdruck aus: "Kohle und Erz", Jg. 1929 und 1930).

____________, "Die geschichtliche Entwicklung des Steinkohlenbergbaus in der heutigen Provinz Westfalen bis zum Jahre 1865," Kohle und Erz (1930).

Mine management; Acta, betreffend die Einsendungen der jährlichen Verwaltungsberichte. Fortgesetzt unter dem Titel: Verwaltungsberichte. Betriebs-und Haushaltssachen. (Generalberichte des Bochumer und des Essen-Werdenschen Bergamts), vol. 2-7, Fotokopie. Bergbaubücherei Nr. 8415 a. Zitiert: Gen.-Ber. 3. Bd.

Ministerial Order of May 10, 1853, *Z.f.B.H.S.W.*, 1853.

Müller, L., *Über die gesetzliche Notwendigkeit und die Vorteile des gemeinsamen Kohlenverkaufs zu amtlich festzustellenden Preisen auf den Ruhrdebitszechen im Essen-Werdenschen, im Gegensatz zu der bisherigen Naturaltheilung und der Veräusserung zu willkürlichen Preisen*, Bonn: 1838.

______________, *Nachtrag und Aktenstücke zu der gedruckten Mitteilung des Gewerken Ludwig Müller in Bonn an seine Mitgewerken, in Betreff des gemeinsamen Kohlenverkaufs zu amtlich festzustellenden Preisen auf den Ruhrdebitszechen im Essen-Werdenschen*, Bonn: 1840.

Müller-Erzbach, R., *Das Bergrecht Preussens und des weiteren Deutschlands*, Stuttgart: 1916.

Münster, Staatsarchiv
(ordered by principal subject)

Anträge der Gewerken auf Abänderung verschiedener Bergwerks Ver fassungs-Verwaltungsgrundsätze 1834-1848, O.B.D. B Gr. 2, Nr. 51.

Allgemeine Berechtsamangelegenheiten (deals with searching permits, claims, and concessions and the legal problems that arose) 1832-1857, O.B.D. B Gr. 68.

Bericht vom 22 Februar 1843 O.B.D.-84/17.

B.A. Bochum, Bericht vom 19. 7. 1816 O.B.D.-37/12.

Entwurf, Revision, und Einführung der neuen Berggesetze, 1826-1828, O.B.D. B Gr. 2, Nr. 36, 37.

Entwurf, Revision, Einführung der neuen Berggesetze 1836-1843, O.B.D. B Gr. 2, Nr. 50.

Gutachten verschiedener Beamten über den Entwurf des neuen Berggesetze 1837... (printed), O.B.D. B Gr. 2, Nr. 50a.

Einführung des Preussischen Bergrechts in den Rhein Provinzen 1816-1821, O.B.D. B Gr. 2, Nr. 46.

Organisation der Bergverwaltung O.B.D. B Gr. 3.

Dampfmaschinen auf Gruben O.B.D. B Gr. 86 (documents Nrs. 1, 2, 3, 17 cover the period 1816-1867, giving among other the number of steam engines in the Ruhr).

Debit und Bergwerksproduktion, Ein-und Ausfuhrzölle... 1841-1857, O.B.D. B Gr. 84, Nr. 17.

Akta, Gewerken von Kunstwerk...Wider den Bergfiskus, Causa I, wegen verweigerten Entrichtung, Zahlung der Zehntgefälle und Freikuxgelder, 1832-1836 O.B.D. B Gr. 55, K.

Prozess gegen Franz Haniel wegen Verweigerter Zahlung des Bergzehnten, 1848-1850, O.B.D. B Gr. 10, Nr. 16, A 157.

Steinkohlenhandel nach Holland...1812-1838, O.B.D. B Gr. 14, Nr. 19, A 466.

Hütten, Hammer, und andere mit dem Bergbau zusammenhängende gewerblichen Anlagen, 1816-1857, O.B.D. B Gr. 3, Nr. e.

Akta, in Sachen des Freiherrn von Schell und Consortium als Gewerken der Zeche Kunstwerk wider den Bergfiskus wegen der von Klägern in Anspruch genommenen Bergwerksabgaben und Freiheiten, 1828-1832, O.B.D. B Gr. 55, Nr. h.

Instruktion für die Oberschichtmeister, November 25, 1832, O.B.D.-24/4.

Von Oeynhausen an von Milicki, 18 Dez. 1843, O.B.D.-C/42.

Akta, die Protokolle der Plenarsitzungen des III. und IV. Westfälischen Provinziallandtages nebst Anlagen enthaltend, Rep. A III 1, Bd. 9-12.

Akta, Anlegung von Eisenbahnen betreffend, Rep. A II 97.

Akta, Verhandlungen des Ausschusses für Handel und Gewerbe, Rep. A III 8.

Rechnungs-und Kassenwesen 1827-1847, O.B.D. B Gr. 92, Nrs. 13, 22, 23, 24, 25.

Reglement wegen Verkaufs der Steinkohlen ins Land und zur Ruhr, 1805-1820, O.B.D. B Gr. 84, Nr. 64.

Regulativ zur Schleusengelderhebung vom 31.5, 1805, Oberpräsidium Münster; B 972.

Übersicht der Steinkohlen Reviere und Gruben des Märkischen Bergamtsbezirk, O.B.D. B Gr. 3, Nr. c.

Akta, Gewerkschaft der Zeche Vereinigung ca. den Bergfiskus wegen verweigerten Zahlung der Zehntgefälle und Freikuxgelder, 1833-1839, O.B.D. B Gr. 55, m.

Patent wegen Verwaltung des Bergragals in...Essen-Werden, 12 April 1803, O.B.D.-522.

Widersetzlichkeiten verschiedener Gewerken gegen die im Jahre 1803 im Essen-Werden ein geführten Preussischen Bergwerks Regal, O.B.D. B Gr. 3, Nr. 11.

Neff, John U., "The mining industry in medieval period," _The Cambridge Economic History of Europe_, Cambridge: Cambridge University Press, 1952.

Die Entwicklung des niederrheinisch-westfälischen Steinkohlen-Bergbaues in der zweiten Hälfte des 19. Jahrhunderts. Herausgeg v. Verein für die bergbaulichen Interessen im Oberbergamtsbezirk Dortmund in Gemeinschaft mit der Westfälischen Berggewerkschaftskasse und dem Rheinisch-Westfälischen Kohlen-Syndikat, Bd. X, XII, Berlin: 1902-1905.

Nordmann, H., "Die ältere preussische Eisenbahngeschichte," _Abhandlungen, Deutsche Akademie der Wissenschaften aus Berlin: Mathematisch- naturwissenschaftliche Klasse_, 1948.

Oberste-Brink, K., "Die Vierung und ewige Teufe der Längenfelder nach dem preussischen Gesetz vom 1. Juli 1821," _Glückauf_, Jg. (1941).

"Bericht über den Betrieb und die Haushaltungsverhältnisse des Phönix , Actien-Gesellschaft für Bergbau und Hüttenbetrieb. Erstattet von dem Generaldirector derselben Charles Detillieux in der dritten Generalversammlung zu Cöln, am 27. Oct. 1856," Berg- und hüttenmännische Zeitung, 15. Jahrg. Freiberg 1856.

Paulus, Otto, Die Stellung des Rheinlandes zur Schutzidee in der Preussischen Wirthschafts-und Sozialpolitik von 1815 bis um die Mitte des 19. Jahrhunderts. (Diss. Köln, 1927).

Pieper, Wilhelm, Ulrich Ruelein von Calw (kalbe) und sein Bergbüchlein, Berlin: Akademie Verlag, 1955.

Poschinger, H., von, Bankwesen und Bankpolitik in Preussen, 3 vol., Berlin: 1878-79.

Pertz, Georg, H., Das Leben des Ministers Freiherrn vom Stein, Berlin: 1850-55, 6 vol.

Pounds, Norman, The Ruhr, Bloomington: Indiana Press, 1952.

______________, The Upper Silesian Industrial Region, Indiana University Publication, Slavic and East European Series, Vol. II (1958).

Pounds, Norman J.G., and Parker, William. Coal and Steel in Western Europe. Bloomington, Indiana: 1957.

"Produktion und Absatz der Ruhrkohlen." Der Berggeist, Jahrg. 1857 und 1858.

Preise der Steinkohlen in dem Distrikte des kgl. Essen-Werdenschen Bergamtes. Aufgestellt vom kgl. Oberbergamt für die westfälischen Provinzen, 1833-1835, 1846, 1851. (Bergbaubücherei Essen, Nr. Fi 141 b).

Preise der Steinkohlen in dem Distrikt des kgl. Märkischen Bergamtes. Aufgestellt vom kgl. Oberbergamt für die westfälischen Provinzen, 1833-1835, 1839 bis 1851. (Bergbaubücherei Essen: Nr. Fi 292 b).

Preussische Jahrbücher, 1858-1935, Berlin.

Prym, A.M., Staatswirtschaft und Privatunternehmung in der Geschichte des Ruhrbergbaus, Essen: 1950.

Reif, Heinrich, "Die modern Strömungen im Bergrechte (Germany)", "Die modernen Strömungen im Bergrechte (Austria)", "Eine Kartellorganization von Staatswegen", "Die preussische Bergesetznovelle vom 18 Junie 1907", Bergrechtliche Blätter (v. 3-7).

Reuss, M., Mitteilungen aus der Geschichte des Königlichen Oberbergamtes zu Dortmund und des niederrheinisch-westfälischen Bergbaues. Festschrift zur feier des hundertjährigen Bestchens des Königlichen Oberbergamtes zu Dortmund am 25. Juni 1892, Berlin: 1892.

Richter, Emil, Die Entwickelung der Verkehrs Grundlagen, Berlin: 1878.

Sax, Emil, Die Verkehrsmittel in Volks-und Staatswissenschaft, Wein: 1878-79.

Schaefer, Gustav, Ursprung and Entwickelung der Verkehrsmittel, Dresden: 1888.

Shareholder Law: "Gesetz über die Verhältnisse der Miteigenthümer eines Bergwerks, 12. Mai, 1851," Z.f.B.H.S.-W., (1854).

Schellhas, Walter, "Die älteste Urkunde des sächsischen Berbaues", Bergbau und Bergleute, Berlin: Adademie Verlag: 1955.

Schnabel, Franz, Deutsche Geschichte im 19. Jahrhundert, Freiburg: 1949.

Schmitthenner, F., Die Main-Weser Eisenbahn, Giessen: 1838.

Schoenbauer, Ernst, "Beiträge zur Geschichte des Bergbaurechts", Münchener Beiträge zur Papyrusforschung und antiken Rechtsgeschichte, München: 1929.

Schwarz, Klaus, Untersuchungen zur Geschichte der deutschen Bergleute im späteren Mitteralter, Berlin: Akademie Verlag, 1958.

Serlo, A., "Beitrag zur Geschichte des schlesischen Bergbaues in den letzten Hundert Jahren," Festschrift zur Feier des hundert-jährigen Bestehens desköniglichen Oberbergamtes zu Breslau, Breslau: 1869.

Serlo, W., "Des Freiherrn vom Stein Verdienste um die Bergwirtschaft", Z.f.B.H.S.-W., (1931).

Simmersbach, "Die Begründung der oberschlesischen Eisenindustrie unter Preussens Königen", Stahl und Eisen Jrg. 1911.

Sombart, Werner, Die deutsche Wokswirtschaft im 19. Jahrhundert und im Anfang des 20. Jahrhundert, Stuttgart, 1954.

Spethmann, Hans, Das Ruhrgebiet, Berlin: 1933.

Spethmann, H., Die ersten Mergelzechen im Ruhrgebiet.

Strotkötter, G., Die Bestrebungen zur Verbesserung der Lippe im 15., 17. und 18. Jahrhundert, Münster: 1895.

__________, Die Lippeschiffahrt im 19. Jahrhundert, Münster: 1896.

Tax Law: "Gesetz über die Besteuerung der Bergwerke, 12. Mai 1851", Z.f.B.H.S.-W., (1854).

Tilly, Richard H., Financial institutions and Industrialization in the Rhineland: 1815-1870, Madison: 1966.

Treptow, Emil, Grundzuege der Bergbaukunde und Aufbereitung, Wien: Spielhagen & Schurch: 1903.

Treue, W., Wirtschaftszustände und Wirtschaftspolitik in Preussen 1815-1825, Stuttgart: 1937.

__________, Wirtschaftsgeschichte der Neuzeit 1700-1960, 1962.

Veit, O., Der Bergbau als staatliche Finanzquelle, Diss. Frankfurt: 1927.

Verein für die bergbaulichen Interessen im Oberbergamtsbezirk Dortmund: Jahresberichte 1858-1863, Essen.

Vowinckel, G., Das Oberbergamt zu Dortmund und der westfälisch-niederrheinische Bergbau 1792-1942, im dienstlichen Auftrag verfasst, Dortmund 1942, Masch.-Schrift, (Standort: Bergbaubücherei Essen, Nr. A 17).

Wagner, Th., Corpus iuris metallici, Leipzig: 1791.

Wagenbreth, Otfried and Hofman, Fritz, Alte Freiberger Bergwerksgebaude und Grubenanlagen, Berlin: Akademie Verlag: 1957.

Waltershausen, Sartorius V., Deutsche Wirtschaftsgeschichte 1815-1914, Jena: 1923.

Weiske, J., Der Bergbau und das Bergregal, Eisleben: 1845.

Weichsel, F., Die Rhein-Weser Eisenbahn A.G., Leipzig: 1838.

Weicks-Glon, Friederich, _Das finanzielle und soziale Wesen der modernen Vekehrsmittel_, Tübingen: 1894.

Westhoff-Schlüter, W., "Geshichte des deutschen Bergrechts," Sonderabdruck aus _Z.F. Bergrecht_ (1909).

Westphalen, H.C., _Die Kriegsfuehrung unter Benutzung der Eisenbahnen und der Kampf um Eisenbahnen_, Leipzig: 1882.

Wilsdorf, Helmut, _Georg Agricula und seine Zeit_, Berlin: VEB Deutscher Verlag der Wissenschaften, ;956.

Wiss, Eduard, _Einenbahn oder Kanal?_ Berlin: 1878.

Zycha, Adolf, _Das Recht des aeltesten deutschen Bergbaues bis 13 Jahrhundert_, Berlin, 1899.

_______________, "Ein altes soziales Arbeiterrecht Deutschlands," _Z.f.B._ (1900).

APPENDIX

The following appendix presents quantative information not used before in any study. This data allows for the first time a more complete analysis of mining and related industries and the Ruhr region's overall economic growth. The data falls into three general categories: (1) Data on individual mining operations; (2) Data on finances and ownership; (3) Data on transportation and technology.

The major sources for the output, income, costs, and profit/losses of individual mining operations are "Monthly Mining Reports, 1829-47" and "General Annual Mining Reports, 1823-43". These monthly reports contain the output and profit or losses for individual operations. The profit was the amount left after the immediate and actual costs of operation. The loss was the amount the owner or shareholder was legally obligated to pay when the income from sales did not cover the monthly cost of operation and taxes. There also some reports for the last quarter of the eighteenth century.

For fiances and ownership in mining and related industries the major source are the "Debt Books". In the Ruhr mining region, any fiancial transaction involving shares in mining and other industrial enterprises was legally required to be entered in a journal or "Debt Book". Borrowers issued financial claims in evidence of debt in form of promisory notes, mortgages, and shares of stock. Consequently, these books document the lending and borrowing activities in the Ruhr region. They are the only available central source on ownership and market value of mining operations prior to 1850's. They provide a host of valuable information including the names of shareholders and their creditors, number of shares and mines, occupation and residence of owners and creditors, and type, amount and occasionally purpose of debt.

Data on transportation and technoloby was gathered from a variety of sources. There is detailed data on Ruhr and Rhine rivers transport, development of iron rails both above and below ground in mining operations, and later, the railroad system. Other data focused on the crucial role of the steam-engine as a power source for transport, actul mining and other industrial operations.

The following are the locations of the major sources available and used in this appendix: Düsseldorf Hauptarchiv and its Zweigarchiv; Dortmund Oberbergamt and the Dortmund Wirtschaftliches Archiv; Münster Staatsarchiv and Münster Landesvermessungsamt; Essen Bergbaubücherei.

The Münster Staatsarchiv is by far the most important place for research on the early Ruhr region. It contains the central mining office records - Oberbergamt Dortmund Akten - included are: (1) The Monthly Mining Reports, 1829-47 for the Mark mining region; (2) General Annual Mining Reports, 1823-43 for the entire Ruhr region; (3) The Debt Books (five volumes, 1780's to 1850's) as major source for ownership and fiances in the Ruhr mining region. The Münster Landesvermessungsamt has an excellent map collection dating back to the late eighteenth century.

The Düsseldorf Hauptarchiv and its branch archive has nineteenth century mining records mostly for the Rhenish part of the Ruhr region as well as other local holdings including the Kleve Kammer Akten, Stift & Kloster Archiv Essen, and Abtei Essen records.

The Dortmund Oberbergamt Archiv has only individual business records for some Ruhr mining operations. The Dortmund Wirtschaftsarchiv has some firm records such as the Harkort-Kamp Firm Archiv. Finally the Essen Bergbaubücherei is a comprehensive mining library.

LIST OF TABLES

1. Average Profit/Losses of Mark Coal Operations, 1830-47

2. Map: Ruhr Mining Districts

3. Annual Profit/Losses for Mark Coal Operations, 1830-47

4. Annual Profit/Losses for Essen Coal Operations, 1836-42

5. Summary: Annual Profit and Losses, 1830-47

6. Chart: Annual Profit and Losses, 1830-47

7. Annual Output for Mark Coal Operations, 1830,36-42,47

8. Annual Output for Essen Coal Operations, 1836-42

9. Averages for Entire Ruhr Coal Operations, 1836-42: Output, Income, Cost, Profit, Taxes

10. Map: Transportation Modes & Location of Mining Operations

11. Mark Region Iron Rails - Above & Below Ground, 1836-48

12. Mark Region Iron Rails - Length & Cost, 1836-48

13. Mark Coal Rails: Length, Cost, Power, Destination, 1848

14. Essen Region Coal Rails: Length & Cost, 1836-40

15. Manufacturers of Steamengines & their Use, 1839

16. Steamengines in Essen Region: Use & Horsepower, 1814-42

17. Price of Steamengines, 1814, 1828, 1836

18. Cost of Operating Steamengines, 1828 & 1836

19. Iron Works Using Steamengines, 1839

20. Textile Operations Using Steamengines, 1839

21. Owners of Mining & other Industria Enterprises and their Creditors: Amount & Year of Debt, Name, Occupation, Residence - 1790' to 1850's

22. Amount & Frequency of Loans - Summary, 1790's to 1850's

23. Chart of Mark Mining Debt

24. Residence of Owners & Creditors: Amount of Debt, 1790's to 1850's

25. Occupation of Mine Owners & Creditors, 1790's to 1850's

26. Mining Office Personnel, 1792

27. Mining Administration Personnel, 1838/39

28. Number of Mining Officials, 1818, 1826, 1838

29. Weights and Measures

TABLE

1830-47, Average Profit/Losses of Mark Coal Mining Operations:
Years in operation, output, Technology, Location

(1) NAME	(2) YEARS OF OPERATION	(3)	(4) PROFIT	(5) OUTPUT	(6) TECHNOLOGY	(7) DISTRICT
1. Gluckauf	(0 0)	18	21567	404900	S I	(3)
2. V. Hasenwinkel & Himmelscrone	(0 0)	18	18693	415200	S R I	(7)
3. Trappe	(0 0)	18	13944	225600	S I	(13)
4. V. Nachtigal	(0 0)	18	12120	260200	S R I	(10)
5. Stock & Scherenberger H.	(0 0)	18	10960	98800	I C	(12)
6. V. General & Erbst	(0 0)	18	10525	354100	R I	(7)
7. Friederica	(0 0)	18	10354	306100	S I	(5)
8. Eintracht	(0 0)	18	8273	183500	R I	(8)
9. Friede	(40 0)	8	7743	268200	R I	(10)
10. Mecklingsbanker E.	(38 0)	10	6726	285100	R I C	(8)
11. Johann Friederich	(0 0)	18	6503	210500	R I	(7)
12. HAMBURG Amts Horde	(0 0)	18	5053	157200		(4)
13. Sieper & Muhler Grube	(0 0)	18	4557	85200	I	(12)
14. V. Tiefbau Morgenstern-Osten	(37 0)	11	4536	215100	R I C	(10)
15. Muhlheimergluck	(0 0)	18	4318	83800	R I	(9)
16. Morgenstern-Osten	(0 40)	12	4188	117600	R I	(10)
17. Eleonora	(31 0)	17	4179	169400	S C	(2)
18. Aufgottgewagt	(0 0)	18	4159	80500	R I	(10)
19. V. Schelle & Haberbank	(0 45)	17	4031	65400	S	(11)
20. V. Altesackberg&Geitling	(0 0)	18	3892	94200	R I	(9)
21. Kirschbaum	(0 0)	13	3729	141900	R I C	(7)
22. Himmelsfursten E.	(0 0)	18	3638	154300	R I	(9)
23. Frischauf	(45 0)	3	3636	132200	I	(4)
24. Stocker Hauptgrube	(0 0)	18	3398	74600	I C	(12)
25. V. Preussischer Adler	(34 0)	13	3390	114400	R I C	(9)
26. Frielinghaus	(0 0)	18	3101	57600	S R	(10)
27. St. Peter	(0 0)	18	2919	59900	S I	(13)
28. Guldene Sonne	(33 45)	11	2716	88600	I C	(5)
29. Aufgottgewagt S.F.	(38 40)	3	2582	0	R	(10)
30. Wilhelmine	(46 47)	2	2449	97800	C	(3)
31. Clarenberg	(0 0)	18	2401	170400	S	(3)
32. Wohlverwahrt	(31 0)	17	2334	92600	R I C	(8)
33. Louise	(0 0)	18	2331	129000	I	(1)

(1)	(2 & 3)		(4)	(5)	(6)	(7)
34. Franziska Tiefbau	(39 0)	9	2160	337700	S R I C	(4)
35. Fundling & Dahlacker	(32 43)	12	2147	148900	S C	(2)
36. Glucksangang III	(0 41)	12	2098	101500		(3)
37. V. Ruhrmannsbank	(0 0)	18	2085	73000	R	(9)
38. Gluckauf & Segen	(35 0)	13	2063	70900	S I C	(10)
39. N. Sackberg & Schrjtenberg	(39 0)	9	2053	31600	R	(10)
40. Morgenstern Westen	(37 39)	3	1953	0	C	(13)
41. Fortuna Westen	(0 0)	18	1909	81300	R I	(8)
42. Dachs & Grevelsloch	(0 0)	18	1886	71700	S I	(4)
43. Einigkeit S.F.	(42 0)	6	1849	89300	R C	(10)
44. Christiane	(41 0)	7	1838	83000		(2)
45. St. Johannes E.S.F.	(37 39)	3	1830	0	R	(3)
46. Hellenbank	(0 44)	15	1826	87700	S	(10)
47. Storcksbank	(30 0)	18	1821	89100	I C	(6)
48. V. Ankunft & Anklang	(0 0)	18	1817	155100	R	(11)
49. Carl Friederich E.	(0 41)	10	1815	85600	R I	(1)
50. Nachtigall	(0 0)	18	1686	39100		(4)
51. Forelle & Alteweib	(0 0)	18	1512	91800	S I	(10)
52. Franziska	(0 0)	18	1500	30400	R	(9)
53. Turteltaube N.	(30 38)	9	1491	31000	R C	(6)
54. Alte Aproche	(0 0)	18	1436	38000	R	(9)
55. Glucksburg	(0 0)	18	1419	88600	R I C	(7)
56. Alte Sackberg	(0 0)	18	1374	39300	R	(11)
57. Christiane & Hilfe Gottes	(32 46)	15	1357	18100	R I C	(3)
58. Gideon	(34 0)	14	1356	45000		(13)
59. St. Moritz	(0 0)	18	1284	35400		(4)
60. Alte Bommerbank E.	(0 0)	18	1264	14100	C	(13)
61. Stucktey	(31 0)	17	1242	43700	I C	(3)
62. Wulfingsburg	(0 0)	18	1233	75100	S I C	(7)
63. Wiendahlsnebenbank	(0 0)	18	1215	45400	I	(12)
64. Glucksonne	(0 38)	9	1205	30200	R	(6)
65. Concordia	(0 0)	18	1187	65700	I C	(6)
66. V. Neue Missgunst	(0 0)	18	1104	118300	R I	(10)
67. Carl Friedrich E.	(32 0)	16	1101	108800	R I C	(13)
68. Reiger	(0 0)	11	1089	40900	R	(4)
69. Rosa	(41 0)	7	1058	58100		(10)
70. Augustus	(0 0)	18	1031	54400	R	(10)
71. Braunschweig N.	(44 0)	4	1013	8200	R C	(10)
72. St. Johannes E.	(0 0)	18	1002	17100	S R I	(10)

	(1)	(2 & 3)		(4)	(5)	(6)	(7)
73.	Stralsund	(0 0)	18	893	119000	R	(10)
74.	Stockerdreckbank	(0 39)	10	873	33300		(12)
75.	V. Bickefeld	(31 0)	17	847	189900	S	(2)
76.	Leveringsbank	(0 38)	9	788	22800		(12)
77.	Franziska Fort.	(37 0)	11	786	38100	R	(4)
78.	Tulipan	(0 0)	18	781	55200		(11)
79.	Neugluck&Stettin	(0 0)	18	754	34000	R	(10)
80.	Alte	(0 37)	8	743	36100		(11)
81.	Hamburg (Witten)	(39 45)	7	738	15100	C	(4)
82.	Gluckauf	(38 0)	10	736	64800	I C	(12)
83.	Wiederlag	(36 0)	12	709	53700	R I C	(10)
84.	Hasenwinkel& H.E.S.F.	(45 0)	3	664	173000	R	(7)
85.	V.Grosse Varst.&Alt.E.	(0 0)	18	641	2800	R I	(9)
86.	Hazard	(31 43)	13	641	26900	R	(10)
87.	Weselbank	(0 30)	1	638	36400	R	(10)
88.	Verinigungs Stol.	(46 0)	2	591	0	R	(9)
89.	Friedrich Wilhelm	(0 0)	18	582	206000	S C	(1)
90.	V. Himmelfahrt	(0 33)	4	570	34600	S R	(3)
91.	Eleonora&Nachtigall	(44 0)	4	567	9100	R I	(10)
92.	V.Anna Cat.&Dickeb.	(0 0)	16	553	53600	R	(7)
93.	Alte.Aproche Schacht.	(0 30)	1	543	24600	R	(9)
94.	Geiltling	(0 0)	18	524	50300		(9)
95.	St. Georg	(0 0)	18	520	14300		(11)
96.	Kirsbaum	(0 33)	4	519	17800	I C	(3)
97.	Juliane	(39 0)	9	515	47800	C	(4)
98.	Cronenebank	(37 0)	11	509	56300		(13)
99.	St.Paul	(43 0)	5	507	17300		(13)
100.	Hermanns Ges.Schiff.	(0 0)	18	490	28300		(11)
101.	Steinbergerbank	(0 0)	16	484	39000	R I C	(4)
102.	Fortuna	(40 46)	7	463	57800		(7)
103.	Hessenbank	(0 32)	3	439	18600		(3)
104.	Glucksanfang	(0 41)	12	437	12400		(5)
105.	Neugluck	(0 0)	18	431	29100		(12)
106.	V.Gluckauf	(0 0)	18	429	50900		(11)
107.	Konigscrone	(35 38)	4	404	25300	C	(6)
108.	Niederhofen	(0 0)	18	384	24900	R I C	(3)
109.	Lucia	(31 0)	17	379	121400	R C	(7)
110.	V.Alexander	(34 41)	8	373	32500	I C	(4)
111.	Clausthal	(36 42)	7	363	8100	I	(4)

(1)	(2 & 3)		(4)	(5)	(6)			(7)
112. Dannenbaum	(0 0)	18	357	49400		I		(5)
113. Schwartze Jung	(0 38)	9	339	49800	R			(8)
114. Hamburg S.F.	(0 30)	1	331	1600	R			(4)
115. V. Dickebackerb.	(41 42)	2	330	10300	R			(7)
116. Turteltaube S.	(0 35)	6	327	30500	R			(10)
117. Buntebank	(0 45)	16	325	31100		I		(3)
118. St. Johannes E.&Friel	(0 32)	3	324	19600	R			(10)
119. Schwarzer Adler	(30 46)	17	320	35900			C	(2)
120. Alte Bommerbank	(0 0)	18	310	41900				(13)
121. Buschbank Nebenbank	(44 0)	4	268	6500				(12)
122. Rabener	(39 45)	7	264	17400	R			(10)
123. St. Joseph	(0 47)	18	260	15700				(11)
124. Timmerbeil	(0 0)	18	228	39700	R			(4)
125. Felicitas	(0 36)	7	179	20800				(2)
126. Neu Gottsgenedich	(0 0)	18	163	25900				(11)
127. Frosch	(0 33)	4	161	16200				(12)
128. Kurzes Ende	(44 0)	4	146	69700			C	(10)
129. Schwarze Pferd	(0 33)	4	145	18900				(2)
130. Oesterbank	(31 37)	7	128	15600				(10)
131. Verlorene Posten	(0 39)	10	125	19300	R			(4)
132. Kaisbergerbank	(36 38)	3	95	24600	R			(10)
133. Harmonie	(38 0)	10	88	49800			C	(11)
134. V. Wildenberg&Vogel.	(0 44)	15	85	31400				(11)
135. Angst & Bange	(39 0)	9	79	10100			C	(3)
136. Charlotte	(0 42)	13	72	15400				(11)
137. Carthauserloch	(0 30)	1	62	0	R			(10)
138. Diederich	(38 0)	10	60	41600			C	(2)
139. Hammerbank	(31 44)	14	48	21600	R	I	C	(6)
140. V. Schierbank	(0 0)	18	45	22600				(11)
141. Portbanker E.	(0 0)	18	41	18400	R			(4)
142. Heinrich S.F.	(46 0)	2	6	14500				(4)
143. Forhe Ansicht	(0 47)	18	4	9800				(2)
144. Dannebaum N.F.	(47 0)	1	-9	20800			C	(5)
145. Elisabeth	(0 39)	8	-21	20800				(2)
146. Friedrich W. Fet.Hen.	(37 37)	1	-35	0				(13)
147. Christian	(0 33)	4	-38	6600				(2)
148. Christine I	(0 33)	4	-44	16000				(3)
149. Hagen Sieperbank	(33 0)	15	-52	46700	R	I	C	(6)

(1)	(2 & 3)	4)	(5)	(6)	(7)
150. Sonnenschein	(35 36) 2	-61	0	R I C	(6)
151. Bergmann	(43 0) 5	-74	22300	R C	(4)
152. Christine &H. Got.S.F.	(42 44) 3	-75	43500	C	(7)
153. Vogelbruch	(34 0)14	-79	25300	R C	(12)
154. Theophilus	(0 37) 8	-90	0	C	(11)
155. Remmelskirchen	(0 38) 9	-99	11500		(11)
156. Cleverbank	(32 32) 1	-102	0	R C	(10)
157. Altemann	(42 42) 1	-111	0	R C	(6)
158. Strauss	(45 45) 1	-120	0	C	(13)
159. Buschbank	(0 0)18	-125	20300		(12)
160 Siebenplaneten	(0 0)16	-128	15000		(4)
161. V.ST.Georg.	(0 0)18	-133	62800		(13)
162. Heinrich	(41 0) 7	-143	31800	I C	(12)
163. Friedlicher Nach.	(37 37) 1	-149	0	C	(4)
164. Johann Mathias	(36 36) 1	-162	0	C	(13)
165. Eintracht S.F.	(42 43) 2	-188	7300	R C	(4)
166. Piepers E.	(42 0) 6	-197	40600	I C	(11)
167. Maximus	(34 45) 11	-208	3200	R	(11)
168. Preussicher Zepter	(39 39) 1	-211	0	C	(8)
169. Govelherdt	(34 34) 1	-212	0	C	(11)
170. V.Caroline&Adler	(30 30) 1	-217	0	C	(10)
171. Krackhausbank	(39 39) 1	-228	0	C	(6)
172. Belle Alliance	(47 0) 1	-234	1500	R C	(11)
173. Jupiter	(45 0) 3	-257	29300	R I C	(2)
174. Oberleveringsbank	(47 47) 1	-269	5800	C	(6)
175. VerlohrnerSohn	(39 42) 4	-277	3100	C	(10)
176. Sandberg	(42 43) 2	-289	1700	C	(10)
177. Adolphus	(31 34) 3	-303	3000	R	(12)
178. Neue Steinkuhl	(37 39) 16	-308	28200	I	(6)
179. Friederich	(33 34) 2	-315	0	R I C	(13)
180. Kurzeggfrbank	(37 39) 3	-317	1600	R C	(6)
181. Nachgedacht	(47 0) 1	-318	18300	C	(6)
182. Bonifacius	(31 34) 4	-325	0	R C	(6)
183. Caroline	(0 0)18	-329	11300		(10)
184. Treue	(0 0)18	-344	20200	R I	(11)
185. Gute Kauf	(37 38) 2	-349	0	R C	(7)
186. Wasser&Wasserne.	(33 34) 2	-363	0	R C	(2)
187. Flachsteich	(0 34) 4	-373	11500		(6)
188. Sengstbank	(0 31) 2	-374	3600		(10)

(1)	(2 & 3)	(4)	(5)	(6)	(7)
189. Margarethe	(0 43) 14	-379	5300	C	(2)
190. Heinrich Wilhelm	(33 43) 11	-386	15800	C	(11)
191. Dreckbank	(0 32) 3	-396	6900		(12)
192. V.Louise	(44 0) 4	-415	300	C	(1)
193. June Witwe	(37 38) 2	-424	0	C	(3)
194. Schultenkamperbank	(38 39) 2	-440	0	C	(1)
195. Catharine	(34 45) 12	-468	400	R	(9)
196. Ferner Gluck	(39 40) 2	-491	0	C	(11)
197. Feigenbaum	(38 44) 7	-494	11800	C	(11)
198. Schligge	(44 0) 4	-497	10000	C	(3)
199. Lustig	(31 31) 1	-505	0	C	(13)
200. Kleingluck	(40 42) 3	-505	5500	C	(12)
201. V.Friedrich Wilhelm	(33 44) 12	-520	5500	R C	(10)
202. Gibralter E.	(31 0) 17	-527	44000	R I C	(6)
203. Friederich	(31 0) 17	-539	10500	C	(6)
204. Glucksfortgang	(39 40) 2	-544	0	R C	(10)
205. Geduld	(37 43) 7	-550	28400	C	(11)
206. Johannes E.	(0 37) 8	-575	24200	R	(4)
207. Hortensia	(41 44) 4	-578	0	C	(6)
208. Altemann	(33 43) 11	-596	35100	R I C	(6)
209. Am Schwaben	(43 44) 2	-638	0	C	(3)
210. Halter	(37 0) 11	-644	100	I C	(6)
211. Lowe	(0 0) 18	-652	11000		(13)
212. Iduna	(42 0) 6	-671	51800	R C	(7)
213. V.Burderschaft	(30 33) 4	-681	0	C	(5)
214. Ignatius	(33 42) 10	-731	9500	R I C	(6)
215. Calr WilhelmE.	(45 0) 3	-761	0	R I C	(6)
216. Der Neue Weg.	(40 0) 8	-762	0	R C	(7)
217. Sieben Sohne	(42 42) 1	-788	2300	C	(11)
218. V.Emmanuel	(46 0) 2	-808	900	C	(4)
219. Frischgewagt	(37 0) 11	-825	0	C	(13)
220. St. Mathias	(0 0) 18	-831	26300	R I	(7)
221. Gluckaufs E.Fluge.	(45 46) 2	-842	0	C	(3)
222. Urbanus	(0 0) 18	-870	134800	S R I C	(6)
223. Isabelle	(33 0) 7	-881	28100	C	(6)
224. Gluckaufer E.	(35 0) 3	-895	1100	C	(3)
225. V.Fiedrichsfeld	(30 0) 18	-906	22000	R	(4)
226. Blumenthal	(47 0) 1	-973	500	C	(3)
227. Nicolaus&Victoria	(43 44) 2	-1043	0	C	(2)

(1)	(2 &	3)	(4)	(5)	(6)	(7)
228. Schiffsruder	(32 36)	5	-1088	700	R C	(6)
229. Einigkeit	(34 0)	14	-1118	46600	R I C	(8)
230. Heller Mittag	(45 0)	3	-1161	400	C	( )
231. Papenbank	(45 0)	4	-1255	1600	R I C	(7)
232. Julius Philip E.	(39 0)	9	-1289	7400	R I C	(6)
233. Elias E.	(31 0)	17	-1294	22000	I C	(11)
234. Martha	(46 0)	2	-1360	800	R C	(10)
235. Helena	(31 0)	17	-1525	8600	I C	(4)
236. V.Charlotte	(35 0)	13	-1654	204700	S R I C	(9)
237. Tiefbau Muldenpartie	(34 35)	2	-1733	0	C	(4)
238. V. Konigscrone	(38 40)	3	-1755	0	C	(6)
239. Wecklenbank	(33 35)	3	-1820	0	R C	(8)
240. Horster E.	(36 0)	12	-2043	0	R I C	(8)
241. Louisengluck	(0 0)	18	-2077	133500	S R I C	(10)
242. Aufgottgewagt Tiefb.	(36 37)	2	-2268	24300	R C	(10)
243. Unverhofft&Freie Vog.	(43 0)	5	-3048	34300	C	(2)
244. St.JohannesE.P.Fre T.	(32 0)	16	-3209	0	S R I C	(10)
245. Egmonts E.	(45 47)	3	-3213	10700	C	(6)
246. Tiefer Romberger E.	(0 42)	12	-3351	0	C	(2)
247. V. Trappe	(44 0)	4	-3499	0	C	(13)
248. V. Maria Anna&Stein.	(43 0)	5	-3793	364000	C	(8)
249. Dachs.Grevel.&Dreckb.	(42 42)	1	-4565	0	C	(12)
250. Schurbank&Charlotteb.	(0 0	18	-4876	148500	C	(2)
251. V. Engelsburg	(34 0)	14	-5039	260700	C	(5)
252. Carl Wilhelm	(43 0)	5	-5113	51300	C	(6)
253. Dreckbanker E.	(41 0)	7	-5523	9200	C	(12)
254. V.Nacht.&Aufgott.T.	(40 0)	8	-6716	230500	C	(10)
255. Crone	(47 0)	1	-7108	2700	C	(3)
256. V.Nachtigall Tiefbau	(32 0)	16	-8174	0	C	(10)
257. Gluckauf Tiefbau	(41 0)	7	-8622	21500	C	(3)
258. V. Prasident	(42 0)	6	-12009	319500	C	(5)
259. Louise Tiefbau	(43 0)	5	-14867	187600	C	(1)

*Explanations:
(1.) Name of operations, (2) Years of Operations between 1830-47,(4) Average Loss but by far most losses are really "Cost of Entry" i.e. investment necessary before an operations shows its first profit-6C stands for this amount;(5) Average output based on years: 1830,36,42,47; (6) Technology; S-Steamengines, I-Iron rails,C-Cost of Entry, see (4), R-Operation having access to Ruhr River; (7) Location-by mining district.

Source: Monthly Mining Reports

EXPLANATION

Name of Mining Operation

As part of the name the following abreviations may be included:
E. = Erbstollen - a drainage tunnel with special legal rights for coal mining.
V. = Vereinigte - consolidated operation or mining fields
S.F. = Streitiges Feld - disputed field
T. = Tiefbau - operation with a shaft

Years of Operation's Existence between 1830-47

An operation in existence for the entire period, 1830-47 will have "18" indicated, for operations in existence less than the full 18 years, dates of entry and/or exit are shown.

Average Annual Profit/Losses, 1830-47

The average profit or loss per year is the sum of all the annual profits and losses divided by the number of years in operation between 1830 and 1847. In turn, the annual profit or loss is based on the monthly profit and loss.

Cost of Entry - C

By the 1830's as mining moved northward and as the coal near the Ruhr river was exhausted, "deep mining" became necessary. It demanded construction of shafts to exploit the deeper layers solve the water problems by introducing steam-engines for both pumping of water and lifting of coal. All these factors were reflected in the cost of entry. This cost may be calculated by adding the losses for the number of years between the initial year of entry and the year for which the operation would show its first annual profit.

RUHR MINING DISTRICTS IN FIRST HALF OF 19th CENTURY

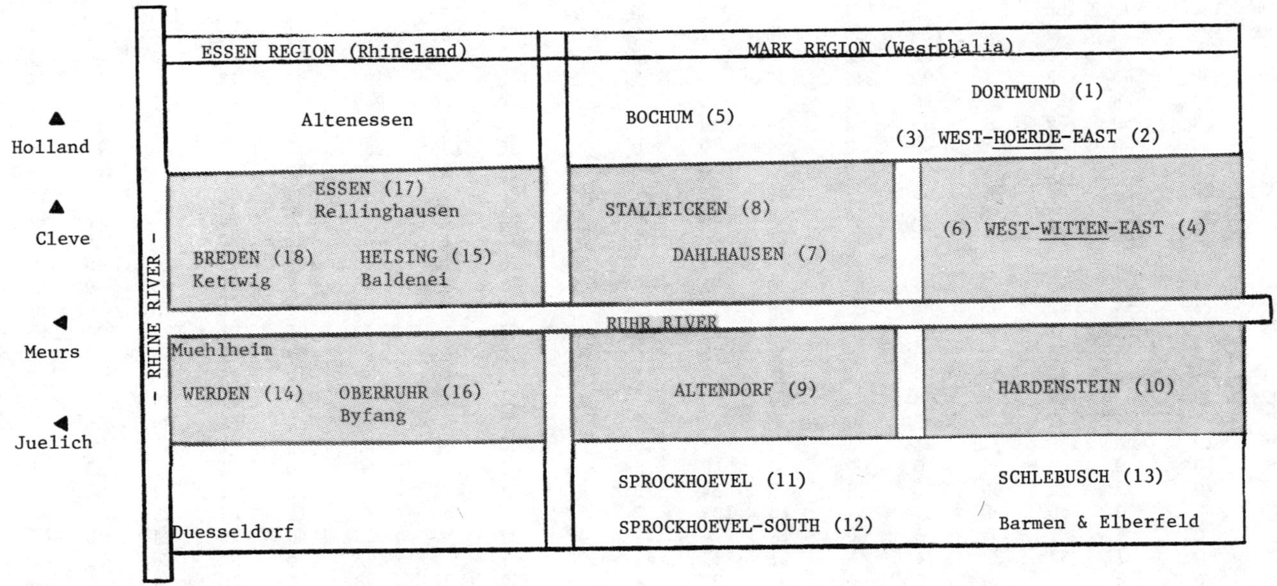

Operations in shaded area have direct access to Ruhr River or its tributaries.

TABLE

1830-47, Annual Profit/Losses for All Mark Coal Operations (100 Rth.)
A. Operations Averaging a Profit, 1830-47
(Arranged by size of Average Profit)*

	AVERAGE	1830	1831	1832	1833	1834	1835	1836	1837	1838	1839	1840	1841	1842	1843	1844	1845	1846	1847
1.	21567	214	173	192	144	158	182	195	181	199	173	166	225	267	308	374	302	253	300
2.	18693	84	119	1	180	187	230	240	226	349	383	267	279	196	228	145	92	161	7
3.	13944	140	71	156	139	118	167	159	131	168	92	104	160	143	154	194	166	129	171
4.	12120	29	13	37	69	139	230	166	388	225	15	77	122	205	169	134	136	144	86
5.	10960	20	32	69	31	-72	122	92	122	282	222	158	186	98	190	80	98	66	99
6.	10525	90	83	92	120	93	140	140	136	121	115	142	115	124	97	77	107	13	67
7.	10354	40	28	66	13	-6	0	79	119	157	183	179	146	242	206	164	99	77	70
8.	8273	34	58	79	56	65	54	73	114	241	146	174	120	2	68	128	33	32	12
9.	7743	0	0	0	0	0	0	0	0	0	0	38	155	132	157	69	20	23	29
10.	6726	0	0	0	0	0	0	0	0	-4	39	88	29	88	57	101	104	6	76
11.	6503	29	20	48	35	62	89	72	68	81	107	90	96	81	77	72	66	0	61
12.	5053	32	33	27	40	50	54	44	54	54	47	52	54	73	76	79	56	77	46
13.	4557	52	48	99	53	76	94	27	65	57	55	34	16	45	42	30	30	8	17
14.	4536	0	0	0	0	0	0	0	-9	-28	1	28	117	149	106	71	14	45	6
15.	4318	73	124	167	75	40	13	20	71	99	25	18	35	-3	7	11	-5	9	-1
16.	4188	73	64	40	48	84	80	39	15	19	11	-1	0	0	0	0	0	0	0
17.	4179	0	-6	-1	-8	-8	-25	-16	-69	-62	-3	60	116	148	105	91	122	126	140
18.	4159	3	8	20	30	26	38	36	31	16	17	52	46	83	93	81	80	77	40
19.	4031	18	-16	45	67	79	74	71	75	58	53	74	-6	-7	19	43	43	0	0
20.	3892	30	47	118	115	80	53	24	90	41	23	32	11	1	13	41	-8	6	-14
21.	3729	-5	4	0	0	0	0	0	0	-13	37	42	64	91	58	42	29	105	-5
22.	3638	19	58	38	11	24	33	11	37	40	16	60	60	53	30	31	69	20	45
23.	3636	0	0	0	0	0	0	0	0	0	0	0	0	0	0	0	40	33	36
24.	3398	79	7	28	49	30	28	58	54	79	87	71	71	31	18	33	13	-52	-26
25.	3390	0	0	0	0	-17	61	40	102	67	53	56	56	-7	13	24	28	-8	-8
26.	3101	0	-4	1	35	-6	68	71	74	102	47	37	37	39	40	16	34	-15	-1
27.	2919	0	7	37	33	54	41	51	55	44	23	7	7	-6	20	31	26	26	22
28.	2716	0	0	0	0	0	-6	11	27	32	33	37	27	48	41	46	9	0	0
29.	2582	0	0	0	0	0	0	0	0	21	43	0	0	0	0	0	0	0	0
30.	2449	0	0	0	0	0	0	0	0	0	0	0	0	0	0	0	0	2	47
31.	2401	-5	-1	2	-15	-59	-18	45	39	5	17	65	65	98	78	94	52	0	30
32.	2334	0	0	10	16	25	6	70	29	36	25	32	32	44	14	23	35	16	33
33.	2331	9	27	17	17	46	23	24	12	14	-2	9	28	53	51	18	40	16	39

AVERAGE	1830	1831	1832	1833	1834	1835	1836	1837	1838	1839	1840	1841	1842	1843	1844	1845	1846	1847
34. 2160	0	0	0	0	0	0	0	0	0	-94	-121	-105	-217	-54	226	215	202	143
35. 2147	0	0	-12	-59	-83	60	45	92	34	80	51	43	55	-5	0	0	0	0
36. 2098	27	26	22	46	31	10	18	18	17	12	1	0	0	0	0	0	0	0
37. 2085	0	3	16	28	26	29	17	32	58	48	19	33	2	9	11	12	8	9
38. 2063	0	0	0	0	0	-23	-24	104	133	90	52	23	-20	-12	-5	-18	0	-33
39. 2053	0	0	0	0	0	0	0	0	0	13	25	22	18	19	34	24	9	0
40. 1953	0	0	0	0	0	0	0	-5	20	25	0	0	0	0	0	0	0	0
41. 1909	12	-3	10	15	21	23	25	18	24	27	16	22	33	28	23	10	4	15
42. 1886	0	-6	2	1	14	7	-7	-5	-24	3	40	45	47	59	39	28	38	
43. 1849	0	0	0	0	0	0	0	0	0	0	0	0	-5	12	23	15	19	47
44. 1838	0	0	0	0	0	0	0	0	0	0	0	9	36	14	13	14	24	20
45. 1830	0	0	0	0	0	0	0	12	25	7	0	0	0	0	0	0	0	0
46. 1826	37	37	47	18	28	0	10	21	15	-1	14	-4	16	0	21	0	0	0
47. 1821	-1	-1	-3	-1	-8	0	0	0	25	19	6	40	52	40	76	24	12	18
48. 1817	18	10	23	22	16	16	7	25	25	18	26	22	15	12	24	28	17	3
49. 1815	4	37	61	52	24	8	1	0	0	0	3	-3	0	0	0	0	0	0
50. 1686	27	23	18	31	17	8	30	59	22	23	22	9	20	10	7	-8	0	-9
51. 1512	-1	14	34	23	12	2	2	2	11	21	17	29	18	8	44	35	14	-7
52. 1500	10	12	16	17	23	25	12	5	18	24	13	17	10	8	16	15	18	16
53. 1491	-5	13	15	19	19	16	25	21	12	0	0	0	0	0	0	0	0	0
54. 1436	14	27	24	28	-3	8	18	-2	5	38	8	40	1	-5	17	23	5	11
55. 1419	-4	-3	-4	-2	15	28	17	28	26	8	1	23	30	28	45	25	17	12
56. 1374	18	36	21	27	23	16	19	34	28	13	6	7	-5	2	5	-1	1	-7
57. 1357	0	0	-5	-34	-11	50	21	59	59	53	23	6	-14	-7	-2	4	0	0
58. 1356	0	0	0	0	16	13	17	33	18	22	19	5	6	6	7	13	11	5
59. 1284	16	9	16	11	17	3	6	7	20	10	5	8	19	23	23	20	21	5
60. 1264	-3	-9	-11	-7	-1	-5	-1	3	-2	-5	1	5	7	9	8	8	13	11
61. 1242	0	-4	-7	13	17	19	14	27	15	14	14	11	14	11	12	10	17	15
62. 1233	0	3	20	11	9	3	4	-1	-2	-23	3	-2	17	37	32	40	26	33
63. 1215	13	12	10	11	11	8	7	16	15	11	5	10	16	8	9	42	29	-5
64. 1205	28	18	28	21	-19	-12	24	19	3	0	0	0	0	0	0	0	0	0
65. 1187	47	20	-1	4	-28	-13	-10	-12	-23	14	21	49	51	44	33	17	13	-28
66. 1104	-26	-8	7	10	-2	-13	-27	88	55	41	25	35	17	26	-4	-7	-13	16
67. 1101	0	0	-8	3	5	5	10	21	-2	36	-19	31	24	7	2	4	3	15
68. 1089	15	36	31	9	-2	4	0	0	0	0	0	0	0	0	-1	-5	1	29
69. 1058	0	0	0	0	0	0	0	0	0	0	0	1	5	4	18	17	17	12
70. 1031	13	10	13	5	16	13	5	4	12	9	10	18	12	13	11	10	7	7

	AVERAGE	1830	1831	1832	1833	1834	1835	1836	1837	1838	1839	1840	1841	1842	1843	1844	1845	1846	1847
71.	1013	0	0	0	0	0	0	0	0	0	0	0	0	0	0	-2	19	0	3
72.	1002	4	9	24	29	22	38	33	8	2	27	8	19	0	7	-16	4	0	-16
73.	893	20	20	14	7	28	21	5	10	10	5	-7	-2	0	-6	4	5	0	0
74.	873	21	19	16	5	4	3	2	1	3	1	0	0	0	0	0	0	0	0
75.	847	0	-4	-46	-59	-114	-89	-36	-5	54	64	68	-31	52	102	141	75	21	-36
76.	788	17	5	21	-1	5	13	7	1	1	3	0	0	0	0	0	0	0	0
77.	786	0	0	0	0	0	0	0	0	18	13	15	12	15	7	2	3	2	1
78.	781	10	13	9	6	10	13	13	10	12	9	8	2	1	3	5	-2	4	19
79.	754	1	10	17	11	10	10	9	5	4	8	10	11	9	2	6	4	1	1
80.	743	7	12	12	10	0	5	12	8	0	0	0	0	0	0	0	0	0	0
81.	738	0	0	0	0	0	0	0	0	0	-9	14	18	12	8	5	2	0	0
82.	736	0	0	0	0	0	0	0	0	0	-30	-17	30	18	-3	0	42	21	18
83.	709	0	0	0	0	0	0	15	-58	37	5	5	11	31	18	14	0	-1	-1
84.	664	0	0	0	0	0	0	0	0	0	0	0	0	0	0	0	3	11	6
85.	641	-4	-7	8	-2	-6	-5	-8	17	12	28	24	30	6	2	0	11	8	5
86.	641	0	2	10	9	6	12	8	11	7	2	2	4	10	0	0	0	0	0
87.	638	9	0	0	0	0	0	0	0	0	0	0	0	0	0	0	0	0	0
88.	591	0	0	0	0	0	0	0	0	0	0	0	0	0	0	0	0	3	9
89.	582	10	26	42	36	65	77	65	58	43	-41	24	-20	-89	-46	-16	-14	-37	-56
90.	570	4	8	13	2	0	0	0	0	0	0	0	0	0	0	0	0	0	0
91.	567	0	0	0	0	0	0	0	0	0	0	0	0	0	0	10	0	8	5
92.	553	19	13	12	-5	0	17	6	0	10	0	0	2	4	1	-2	7	12	2
93.	543	5	0	0	0	0	0	0	0	0	0	0	0	0	0	0	0	0	0
94.	524	15	17	29	2	16	-1	9	-1	6	-5	1	3	1	3	4	2	2	2
95.	520	18	17	9	12	-1	-2	4	6	4	5	2	-12	-3	5	8	8	7	9
96.	519	4	11	6	2	0	0	0	0	0	0	0	0	0	0	0	0	0	0
97.	515	0	0	0	0	0	0	0	0	0	-11	-17	-4	-4	3	22	19	14	24
98.	509	0	0	0	0	0	0	0	-3	2	0	11	12	8	9	8	3	4	1
99.	507	0	0	0	0	0	0	0	0	0	0	0	0	0	2	2	7	8	6
100.	490	2	1	1	3	3	6	5	3	3	5	6	4	6	6	3	4	12	19
101.	484	8	7	11	11	11	8	11	10	4	-10	-8	0	0	0	0	-1	1	3
102.	463	0	0	0	0	0	0	0	0	0	0	-4	-21	2	18	28	12	0	0
103.	439	5	2	4	0	0	0	0	0	0	0	0	0	0	0	0	0	0	0
104.	437	7	3	3	4	2	3	3	7	7	0	2	1	0	0	0	0	0	0
105.	431	14	13	8	3	1	4	4	5	10	0	5	7	-1	5	4	-5	1	-6
106.	429	0	-4	2	4	-1	19	32	19	15	-25	11	11	4	2	0	2	-7	-3
107.	404	0	0	0	0	0	-4	8	13	-1	0	0	0	0	0	0	0	0	0

	AVERAGE	1830	1831	1832	1833	1834	1835	1836	1837	1838	1839	1840	1841	1842	1843	1844	1845	1846	1847
108.	384	0	4	14	22	6	2	-7	-1	0	11	9	5	4	0	0	-1	0	0
109.	379	0	-7	-21	12	3	10	6	-9	-21	-68	31	56	25	30	8	17	-9	2
110.	373	0	0	0	0	-2	0	4	4	6	10	6	2	0	0	0	0	0	0
111.	363	0	0	0	0	0	0	-6	4	17	-5	11	8	4	0	0	0	0	0
112.	357	14	10	5	-14	-7	-4	-3	4	25	4	11	-6	15	20	8	-3	-8	-4
113.	339	3	5	4	4	8	5	-8	8	1	0	0	0	0	0	0	0	0	0
114.	331	0	0	0	0	0	0	0	0	0	0	0	3	3	0	0	0	0	0
115.	330	1	0	0	0	0	0	0	0	0	0	0	0	0	0	0	0	0	0
116.	327	14	2	0	-2	-3	3	0	0	0	0	0	0	0	0	0	0	0	0
117.	325	-7	6	4	5	1	-1	0	-5	-5	4	4	14	12	5	10	3	0	0
118.	324	6	12	-8	0	0	0	0	0	0	0	0	0	0	0	0	0	0	0
119.	320	-6	-12	14	13	2	3	1	4	4	3	2	0	-2	-1	18	14	-1	0
120.	310	-1	1	-1	-7	-7	-2	18	18	13	11	8	2	-3	0	-6	1	4	4
121.	268	0	0	0	0	0	0	0	0	0	0	0	0	0	0	2	5	1	2
122.	261	0	0	0	0	0	0	0	0	0	0	2	0	1	4	4	7	0	0
123.	260	-2	5	12	8	4	3	8	5	2	0	1	4	-2	1	1	0	-2	-1
124.	228	0	1	-2	2	2	4	0	2	2	2	2	-2	0	0	-3	8	12	7
125.	179	7	-2	-2	-7	5	4	0	0	0	0	0	0	0	0	0	0	0	0
126.	163	0	1	-2	6	2	10	1	1	1	-2	-1	2	8	0	-1	3	-1	-1
127.	161	1	3	2	0	0	0	0	0	0	0	0	0	0	0	0	0	0	0
128.	146	0	0	0	0	0	0	0	0	0	0	0	0	0	0	-6	-10	10	12
129.	145	-5	-1	8	-2	0	0	0	0	0	0	0	0	0	0	0	0	0	0
130.	128	0	4	2	3	0	0	-1	3	0	0	0	0	0	0	0	0	0	C
131.	125	0	7	2	0	0	0	0	-3	-4	3	0	0	0	0	0	0	0	0
132.	95	0	0	0	0	0	0	-11	-11	3	0	0	0	0	0	0	0	0	0
133.	88	0	0	0	0	0	0	0	0	-11	-12	-8	-8	-8	-2	9	0	25	25
134.	85	8	3	-3	-2	0	0	0	-1	-3	0	-1	-3	3	2	6	0	0	0
135.	79	0	0	0	0	0	0	0	0	0	-3	-1	5	2	3	0	1	1	0
136.	72	-8	-3	-5	-18	-1	5	8	8	5	7	1	2	0	0	0	0	0	0
137.	62	1	0	0	0	0	0	0	0	0	0	0	0	0	0	0	0	0	0
138.	60	0	0	0	0	0	0	0	0	-5	-8	-2	-1	-4	0	0	0	0	0
139.	48	0	-5	-8	0	-1	-3	8	2	-3	3	3	9	2	-6	5	0	0	0
140.	45	-1	-2	4	1	2	6	4	-1	-9	-4	-2	0	4	5	2	1	-3	1
141.	41	-10	-12	-10	-1	4	6	4	3	-1	2	2	1	3	1	1	2	-1	-1
142.	6	0	0	0	0	0	0	0	0	0	0	0	0	0	0	0	0	0	0
143.	4	-1	-3	0	0	-1	-1	-4	-1	-2	0	3	-1	-4	6	7	7	-1	-2

B. Operations Averaging a Loss, 1830-47

	AVERAGE	1830	1831	1832	1833	1834	1835	1836	1837	1838	1839	1840	1841	1842	1843	1844	1845	1846	1847
144.	-9	0	0	0	0	0	0	0	0	0	0	0	0	0	0	0	0	0	0
145.	-21	1	-5	-4	-3	1	1	1	4	2	1	0	0	0	0	0	0	0	0
146.	-35	0	0	0	0	0	0	0	0	0	0	0	0	0	0	0	0	0	0
147.	-38	1	-4	0	0	3	0	0	0	0	0	0	0	0	0	0	0	0	0
148.	-44	9	-8	-2	3	-1	0	0	0	0	0	0	0	0	0	0	0	0	0
149.	-52	0	0	0	-17	-14	-7	-12	29	-12	-11	-1	1	7	9	10	9	2	0
150.	-61	0	0	0	0	0	0	1	0	0	0	0	0	0	0	0	0	0	0
151.	-74	0	0	0	0	0	0	0	0	0	0	0	0	0	-4	-5	1	1	2
152.	-75	0	0	0	0	0	0	0	0	0	0	0	0	-10	2	5	0	0	0
153.	-79	0	0	0	0	-6	-8	2	1	7	3	9	7	-3	0	0	-13	-7	-3
154.	-90	-3	0	-2	-1	-2	0	0	0	0	0	0	0	0	0	0	0	0	0
155.	-99	-7	-7	-1	2	4	-1	6	-2	-1	0	0	0	0	0	0	0	0	0
156.	-102	0	0	-1	0	0	0	0	0	0	0	0	0	0	0	0	0	0	0
157.	-111	0	0	0	0	0	0	0	0	0	0	0	0	0	-1	-1	0	0	0
158.	-120	0	0	0	0	0	0	0	0	0	0	0	0	0	0	0	-1	0	0
159.	-125	-2	-3	-3	-7	-4	-2	-3	3	-2	14	2	1	0	-24	-2	3	6	8
160.	-128	-1	0	-1	-1	-3	-1	-2	-1	0	-1	1	1	0	0	0	-5	-4	-5
161.	-133	0	0	1	-1	1	0	0	3	0	-2	-1	-4	0	3	2	0	-8	-15
162.	-143	0	0	0	0	0	0	0	0	0	0	0	-5	-1	-20	-1	8	1	5
163.	-149	0	0	0	0	0	0	0	-1	0	0	0	0	0	0	0	0	0	0
164.	-162	0	0	0	0	0	0	-2	0	0	0	0	0	0	0	0	0	0	0
165.	-188	0	0	0	0	0	0	0	0	0	0	0	0	-4	4	0	0	0	0
166.	-197	0	0	0	0	0	0	0	0	0	0	0	0	-27	-14	10	8	6	4
167.	-208	0	0	0	0	-7	-2	0	-1	-10	-11	-2	-2	7	-2	2	3	0	0
168.	-211	0	0	0	0	0	0	0	0	0	-2	0	0	0	0	0	0	0	0
169.	-212	0	0	0	0	-2	0	0	0	0	0	0	0	0	0	0	0	0	0
170.	-217	-2	0	0	0	0	0	0	0	0	0	0	0	0	0	0	0	0	0
171.	-228	0	0	0	0	0	0	0	0	0	-2	0	0	0	0	0	0	0	0
172.	-234	0	0	0	0	0	0	0	0	0	0	0	0	0	0	0	0	0	-2
173.	-257	0	0	0	0	0	0	0	0	0	0	0	0	0	0	0	-2	-6	-1
174.	-269	0	0	0	0	0	0	0	-3	0	0	0	0	0	0	0	0	0	-3
175.	-277	0	0	0	0	0	0	0	0	0	-2	-5	0	0	0	0	0	0	0
176.	-289	0	0	0	0	0	0	0	0	0	0	0	0	-6	-4	0	0	0	0
177.	-303	0	-5	-1	-1	0	0	0	0	0	0	0	0	0	0	0	0	0	0
178.	-308	0	0	-9	3	-3	-1	5	6	-18	4	-9	4	-2	3	-5	-17	-2	0
179.	-315	0	0	0	-7	1	0	0	0	0	0	0	0	0	0	0	0	0	0
180.	-317	0	0	0	0	0	0	0	-3	-9	2	0	0	0	0	0	0	0	0

AVERAGE	1830	1831	1832	1833	1834	1835	1836	1837	1838	1839	1840	1841	1842	1843	1844	1845	1846	1847
181. -318	0	0	0	0	0	0	0	0	0	0	0	0	0	0	0	0	0	-3
182. -325	0	-2	-5	-3	0	0	0	0	0	0	0	0	0	0	0	0	0	0
183. -329	5	5	6	-1	1	-10	-7	-9	-6	-3	-8	-3	-6	0	0	-1	-10	-13
184. -344	0	2	6	-5	-7	-26	-26	8	9	-27	8	2	1	8	-21	-6	-3	-1
185. -349	0	0	0	0	0	0	0	-7	0	0	0	0	0	0	0	0	0	0
186. -363	0	0	0	-7	12	0	0	0	0	0	0	0	0	0	0	0	0	0
187. -373	0	0	0	-4	0	0	0	0	0	0	0	0	0	0	0	0	0	0
188. -374	0	-4	-4	0	0	0	0	0	0	0	0	0	0	0	0	0	0	0
189. -379	-7	-5	-2	0	-1	-1	-11	-2	1	-1	-2	-4	-11	0	0	0	0	0
190. -386	0	0	0	-4	-11	-8	-3	-2	-5	-1	-1	-5	-1	-2	0	0	0	0
191. -396	-2	-8	-5	0	0	0	0	0	0	0	0	0	0	0	0	0	0	0
192. -415	0	0	0	0	0	0	0	0	0	0	0	0	0	0	0	0	0	0
193. -424	0	0	0	0	0	0	0	0	0	0	0	0	0	0	-4	-3	-1	-8
194. -440	0	0	0	0	0	0	0	-7	-2	0	0	0	0	0	0	0	0	0
195. -468	0	0	0	0	0	0	0	0	-4	-4	0	0	0	0	0	0	0	0
196. -491	0	0	0	0	-2	-3	0	-3	-7	-7	-5	-5	-4	-4	-8	-7	0	0
197. -494	0	0	0	0	0	0	0	0	0	-6	-4	0	0	0	0	0	0	0
198. -497	0	0	0	0	0	0	0	0	-5	-8	-7	-6	-2	-5	-1	0	0	0
199. -505	0	0	0	0	0	0	0	0	0	0	0	0	0	0	-6	-1	-2	-10
200. -505	0	0	-5	0	0	0	0	0	0	0	0	0	0	0	0	0	0	0
201. -520	0	0	0	0	0	0	0	0	-1	0	-3	-13	-3	0	0	0	0	0
202. -527	0	0	0	-8	-4	-2	-3	-1	-6	-8	-7	-7	-6	-6	-4	0	0	0
203. -539	0	-15	-51	-11	-24	-5	0	34	45	-4	-13	-6	-12	14	-7	9	-27	-17
204. -544	0	-3	-8	-8	-13	-7	-2	12	9	-3	2	10	-10	-3	-61	-38	-6	36
205. -550	0	0	0	0	0	0	0	0	0	-2	-8	0	0	0	0	0	0	0
206. -575	0	0	0	0	0	0	0	-12	-11	-9	-1	-7	3	4	0	0	0	0
207. -578	-6	-10	-27	-4	-4	5	-2	3	0	0	0	0	0	0	0	0	0	0
208. -596	0	0	0	0	0	0	0	0	0	0	0	-4	-8	-8	-3	0	0	0
209. -638	0	0	0	-2	-17	-20	-21	25	8	9	4	1	-11	-41	0	0	0	0
210. -644	0	0	0	0	0	0	0	0	0	0	0	0	0	-8	-5	0	0	0
211. -652	0	0	0	0	0	0	0	-4	-4	-5	-5	-8	-6	-7	-8	-8	-6	-6
212. -671	0	0	-2	-4	-8	3	0	-1	-3	2	1	5	1	0	1	1	-3	-2
213. -681	0	0	0	0	0	0	0	0	0	0	0	0	-12	-12	-16	-2	-4	6
214. -731	0	0	-10	-5	0	0	0	0	0	0	0	0	0	0	0	0	0	0
215. -761	0	0	0	-5	-16	-15	-3	2	-20	-16	-1	1	0	0	0	0	0	0
216. -762	0	0	0	0	0	0	0	0	0	0	0	0	0	0	0	-9	-11	-4

	AVERAGE	1830	1831	1832	1833	1834	1835	1836	1837	1838	1839	1840	1841	1842	1843	1844	1845	1846	1847
217.	-788	0	0	0	0	0	0	0	0	0	0	0	0	-8	0	0	0	0	0
218.	-803	0	0	0	0	0	0	0	0	0	0	0	0	0	0	0	-1	-6	-10
219.	-835	0	0	0	0	0	0	0	-4	-11	-13	-12	-10	-8	-5	-8	-8	-8	-3
220.	-831	-4	-4	-10	-14	-13	-17	-17	-11	-17	1	-10	-9	6	-13	0	8	-13	-14
221.	-842	0	0	0	0	0	0	0	0	0	0	0	0	0	0	-1	-7	-10	0
222.	-870	23	21	40	37	25	37	31	04	20	-6	-169	-156	-113	-97	-148	67	114	124
223.	-881	0	0	0	-9	-25	-8	-10	-7	0	0	0	0	0	0	0	0	1	-3
224.	-895	0	0	0	0	0	-2	-7	-11	-8	-11	-23	-23	-17	-5	-1	2	3	-13
225.	-906	0	-2	-7	-7	-10	-8	-11	-17	-28	-43	-10	-5	-13	6	5	-2	-3	0
226.	-973	0	0	0	0	0	0	0	0	0	0	0	0	0	0	0	0	0	-10
227.	-1043	0	0	0	0	0	0	0	0	0	0	0	0	0	-12	-9	0	0	0
228.	-1088	0	0	-18	-20	-14	-2	-1	0	0	0	0	0	0	0	0	0	0	0
229.	-1118	0	0	0	0	-23	-20	-22	-24	-38	-32	-31	-5	-11	-4	27	2	1	24
230.	-1161	0	0	0	0	0	0	0	0	0	0	0	0	0	0	-1	-17	-12	0
231.	-1255	0	0	0	0	0	0	0	0	0	0	0	0	0	0	-2	-13	-8	-27
232.	-1289	0	0	0	0	0	0	0	0	0	-26	-14	-22	-51	-8	-1	7	15	-16
233.	-1294	0	-1	-13	-25	-23	-5	14	-10	-21	-14	-27	-13	-13	-19	-9	-12	-18	-11
234.	-1360	0	0	0	0	0	0	0	0	0	0	0	0	0	0	0	0	-3	-24
235.	-1525	0	-1	-13	-21	-27	-25	-9	-5	-13	-15	-14	-20	-12	-14	-15	-11	-18	-26
236.	-1654	0	0	0	0	0	-120	-73	21	-6	16	-19	-70	-76	83	32	150	69	-11
237.	-1733	0	0	0	0	-24	-11	0	0	0	0	0	0	0	0	0	0	0	0
238.	-1755	0	0	0	0	0	0	0	0	-3	-25	-26	0	0	0	0	0	0	0
239.	-1820	0	0	0	-15	-15	-25	0	0	0	0	0	0	0	0	0	0	0	0
240.	-2043	0	0	0	0	0	0	-33	-27	-36	-30	-30	-29	-23	-29	-29	1	4	17
241.	-2077	31	37	31	21	2	2	-60	-68	63	37	9	-31	-85	-36	2	-125	-142	-60
242.	-2268	0	0	0	0	0	0	11	-55	0	0	0	0	0	0	0	0	0	0
243.	-3048	0	0	0	0	0	0	0	0	0	0	0	0	0	-154	-109	-13	130	-7
244.	-3209	0	0	-4	-10	-97	-97	-48	-72	-45	-50	-2	42	38	34	-17	-31	-53	-8
245.	-3213	0	0	0	0	0	0	0	0	0	0	0	0	0	0	0	-42	-15	-21
246.	-3351	-44	-55	-67	-64	-30	-20	-24	-49	-41	-1	0	0	-13	0	0	0	0	0
247.	-3499	0	0	0	0	0	0	0	0	0	0	0	0	0	0	-9	-50	-30	-51
248.	-3793	0	0	0	0	0	0	0	0	0	0	0	0	0	-75	-64	11	-71	9
249.	-4565	0	0	0	0	0	0	0	0	0	0	0	0	-46	0	0	0	0	0
250.	-4876	45	9	-5	-50	-41	-27	-20	-20	32	70	-4	55	-102	-177	-165	-340	-70	-107
251.	-5039	0	0	0	0	-24	-145	-187	-130	-13	-19	-10	-53	15	38	127	-31	-163	-131
252.	-5113	0	0	0	0	0	0	0	0	0	0	0	0	0	-50	-100	-48	-42	-21
253.	-5523	0	0	0	0	0	0	0	0	0	0	0	-15	-57	-76	-83	-55	-58	-43

AVERAGE	1830	1831	1832	1833	1834	1835	1836	1837	1838	1839	1840	1841	1842	1843	1844	1845	1846	1847
254. -6716	0	0	0	0	0	0	0	0	0	0	-130	-92	-237	-156	-34	-49	80	81
255. -7108	0	0	0	0	0	0	0	0	0	0	0	0	0	0	0	0	0	-71
256. -8174	0	0	-55	-123	-92	-182	-41	-53	-99	-156	-86	-84	-78	-80	-58	-46	-41	-26
257. -8622	0	0	0	0	0	0	0	0	0	0	-24	-97	-86	-98	-91	-92	-88	-112
258.-12009	0	0	0	0	0	0	0	0	0	0	0	0	-22	-235	-194	-174	-42	-53
259.-14867	0	0	0	0	0	0	0	0	0	0	0	0	0	-191	-125	-140	-180	-107

Source: Monthly Mining Reports, 1830-47

TABLE

1836-42, ANNUAL PROFIT/LOSSES FOR ALL ESSEN COAL OPERATIONS (in 100 Rth.)

(Arranged by order of average profit/loss)

	AVERAGE	1836	1837	1838	1839	1840	1841	1842
SAELZER - NEUAK	27960	329	295	289	283	270	239	253
SCHOLERPAD	13723	-154	-27	231	258	366	144	143
KUNSIWERK	10671	-8	82	0	31	151	287	230
HITZBERG & DICKEB.	8290	-2	140	55	63	46	171	109
STEINGATT UNTER MARK	7948	52	38	82	32	109	112	133
GEWALT	6958	-143	-53	51	112	218	151	138
LANGENBRAHM	6904	-90	2	17	100	61	189	208
PORTINGSIEPEN	5287	3	9	13	112	94	-7	145
HUNDSNACKEN	4772	53	52	1	44	54	119	10
SAELZER - NEUAK S.F.	2430	0	0	-10	23	58	11	5
BRUCHKAMP & STEINKNAPPE	2174	14	46	32	-13	4	47	-2
BRAUT - KUPERSWIESE	1844	13	-1	34	19	20	2	32
BALDEMEIER E.	1655	2	1	3	20	29	44	27
SCHWARZE ADLER	1549	-16	14	17	-29	47	53	22
V. GIRONDELLE	1545	1	27	19	-19	30	60	-13
STEINGATT UNTER HEINRICHSFELD	1518	9	30	7	0	0	0	0
BRAUT & NEBENBANK	1453	11	7	19	21	14	0	0
ILANDSBRAUT	1345	-20	23	16	-23	27	41	28
DUVENKAMPSBANK	1246	37	-1	6	-19	52	14	-1
NEUGLUCK S.F.	1124	0	30	3	-1	0	0	0
WOHLGEMUT	1100	-8	30	3	-1	0	0	0
KAMPGESBANK	1049	-12	-3	-6	47	17	26	7
V. FELD	842	8	0	0	0	0	0	0
NIERENBANK	792	-2	-5	-18	8	33	32	-2
PFAFFENBERG	663	0	18	0	0	17	-2	-1
STEINGATT DICKEB. S.	591	11	10	4	0	-6	-1	1
FLOTE	550	-5	23	3	1	0	0	0
V. KUCKS & SCHINKENBANK	526	7	11	4	2	5	3	-1
V. HENRIETTE	401	-161	155	-28	17	165	47	-165

	AVERAGE	1836	1837	1838	1839	1840	1841	1842
PLATZGESBANK H.	357	-4	11	0	0	0	0	0
SCHINKENBANK	350	-2	4	2	6	3	8	0
FORTUNA	207	1	3	0	0	0	0	0
PREUSEN BORBECK SIEPEN	203	-1	4	-28	-15	-13	22	45
SCHMACHTENBANK	171	0	2	2	2	2	2	1
ERBENBANK	168	2	1	2	0	5	0	3
UBERIENKUNFT	160	0	3	-1	0	0	0	0
JACKE	154	4	6	6	2	1	-5	-3
KAHRERWIEDERTOPFE	87	0	-1	2	0	-5	-3	0
HERRENBANK	53	0	1	1	0	0	0	2
PLATZGESBANK NEBENFLOTZ	51	-4	10	-3	-1	3	4	-7
NEUGLUCK	12	7	3	0	-2	2	10	0
SELLERBECK	0	0	0	0	0	0	0	0
WUESCGE	0	0	0	0	0	0	0	0
KLEIN UMSTAND	-30	0	0	1	0	0	0	0
V. HIPPE	-31	5	8	8	-8	-18	4	-1
VEREINIGUNG UNTER STOLLENSOHLE	-74	0	0	-26	-48	-100	-147	-174
NICOLAUS	-145	0	0	0	0	0	0	-1
ERBENKAMPSBANK	-164	-8	3	-4	-4	-3	-4	-3
TROMPETE	-165	2	-2	1	-9	0	7	-6
ANSELMUS E.	-179	-3	-9	0	0	-1	3	-1
NOTTEKAPPE	-253	0	3	-15	-3	-5	6	-1
CATHARINA LEVERMANNSFELD	-304	0	0	0	0	0	3	-9
V. CATHARINA	-341	-4	6	-5	-5	-5	-1	-10
V. HERMANN	-342	0	0	-19	-17	28	17	-15
PRINZFRIEDRICH	-362	0	0	0	0	0	-5	-3
V. HOFFNONG	-370	-1	10	-11	-9	-5	-2	-7
WILHELMINE SCHNEPPE	-498	0	0	0	0	-4	1	0
GROSSE BOVERMANNSWIESE	-412	0	0	0	0	-4	-6	-10
V. PAUL & V. HIPPE ST.	-433	-5	-3	-5	-7	-2	0	0
SCHUTZDELLERBANK								
V. CAPPELLENBANK	-467	-23	25	-23	-15	3	2	-1
IROMPETE UNTER MARK	-472	-1	-1	-4	-9	-18	3	18

	AVERAGE	1836	1837	1838	1839	1840	1841	1842
MUHLE	-488	-5	1	0	0	0	0	0
FRISCHES GLUCK & FRIEDERICH	-495	-5	-4	-9	-1	-1	0	0
BETTELSTRUCKER ST.	-543	-8	-1	-5	-14	-5	0	-2
V. DEIMELSBERGER E.	-565	-3	2	-16	-5	-6	-6	0
MAASBANK	-586	0	-6	-12	-11	-7	-2	-3
NIPSHAGEN	-589	2	0	1	0	0	0	0
BEISENKAMP	-689	-8	1	-11	-3	-4	-8	-16
PAULINER E.	-758	-5	-7	-8	-13	-7	-13	-8
GOLDFINKEL	-824	0	0	0	-7	-10	0	0
STEINGATT	-832	4	-53	-56	44	-27	-33	9
HEIMANNSFELD	-835	0	0	-5	-13	0	-18	-6
PAUL	-934	0	0	0	0	-3	-7	-18
KALSKSIEPEN	-969	-5	-14	-13	-20	2	-25	7
TROTZ	-1112	-15	-8	-14	-17	-2	-2	0
KRONENBERG	-1682	0	0	0	0	0	-15	-19
V. LOUISE	-1853	0	0	0	-6	-50	16	-22
V. HOBEISEN	-4477	-88	-42	-26	-91	-173	-58	-23
AMSEL & SCHNEPPE	-4520	0	0	0	0	0	-9	-34
VEREINIGUNG UBER STOLLENSOHLE	-9904	0	-4	-9	5	5	0	0
KRONENPRINZ V. PREUSSEN	-10409	-173	-182	-135	-171	-167	-53	-16
MONCHHOFFSBANK	-12505	0	0	-4	-43	-189	-128	-143
WOLFSBANK	-19105	0	0	-88	-166	-177	-136	-260
V. HAGENBECK	-19883	0	0	-10	-191	-299	-259	-244
GRAF BEUST	-53094	0	0	0	0	0	0	-531

SOURCE: MONTHLY MINING REPORTS,1830-47
ANNUA MINING REPORTS, 1836-42

TABLE

1830 - 47, Annual Profit and Losses in the Mark Region
(in 1,000 thalers)

Years	Total: Profit	Losses	Net	Ten operations with highest: Profit	Losses
1830	174	18	156	90	13
1831	171	23	148	85	15
1832	224	42	182	111	31
1833	209	75	134	101	56
1834	206	108	98	108	64
1835	256	97	159	141	76
1836	246	83	163	129	56
1837	335	84	251	162	62
1838	368	72	296	198	42
1839	305	87	218	150	57
1840	297	112	185	145	67
1841	341	106	235	163	75
1842	339	160	179	170	114
1843	343	185	158	173	134
1844	374	150	224	172	114
1845	322	147	175	144	112
1846	281	136	145	144	91
1847	250	123	127	129	79
Total	5,041	1,808	3,233	2,515	1,278

Source: Monthly Mining Reports, 1830-47

1830-47, ANNUAL MARK PROFITS AND LOSSES:

QUANTITY (Rth.) AND NUMBER OF OPERATIONS

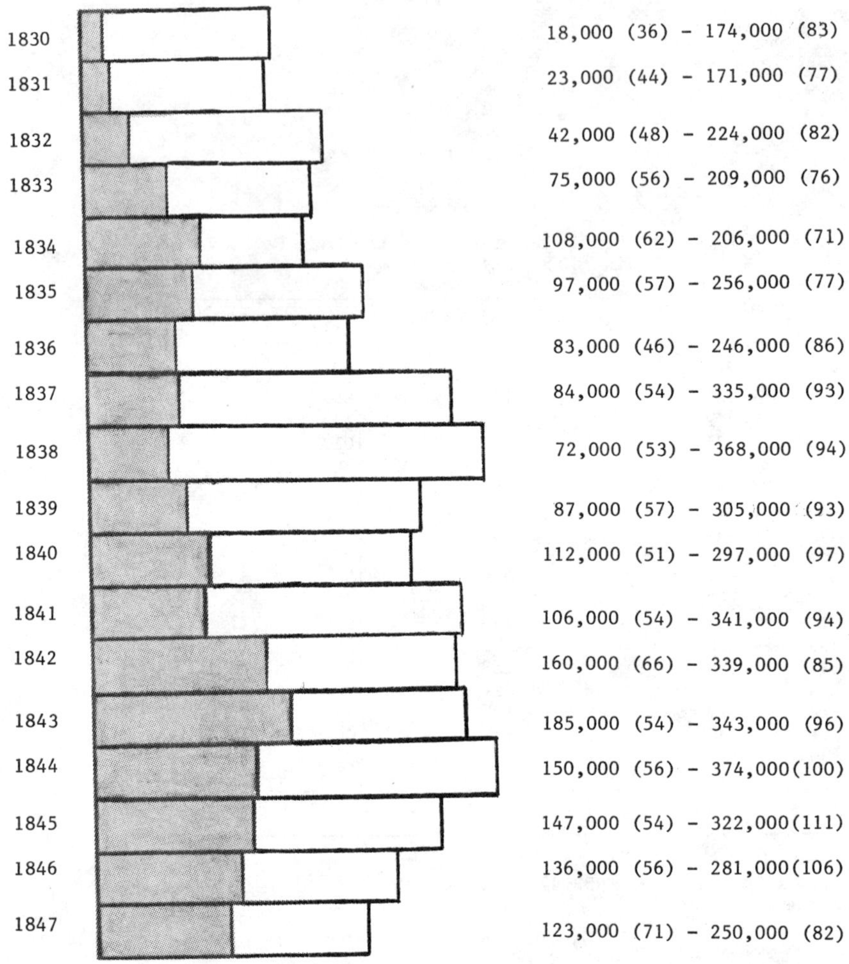

Shaded Area: Losses which include outlays to start operations and for machines and tunneling or shaft work of existing operations

Source: Monthly Mining Reports, 1830-47

TABLE

1830, 36-42, 47 ANNUAL OUTPUT FOR ALL MARK COAL OPERATIONS (100 Scheffel)

(Arranged by average output-Average is based on years 1830, 36, 42, 47 output)

	NAME	YEARS OPERATING 1830-47	AVER. AGE	1830	1836	1837	1838	1839	1840	1841	1842	1847
1.	V.HASENWINKEL&HIMMELSCRONE	18	4152	3795	5073	6568	7291	7342	5050	5087	4101	3642
2.	GLUCKAUF	18	4049	4544	3032	2695	3211	3407	3401	3681	3926	4697
3.	V. MARIA ANNA & STEINBERG B.	5	3640	0	0	0	0	0	0	0	0	3640
4.	V. GENERAL & ERBST.	18	3541	161	3901	5398	6849	6200	5944	5823	4802	5301
5.	UNVERHOFFT & FREIE VOGEL	5	3430	0	0	0	0	0	0	0	0	3430
6.	FRANZISKA TIEFBAU	9	3377	0	0	0	0	0	0	126	2174	4581
7.	V. PRASIDENT	6	3195	0	0	0	0	0	0	0	0	3195
8.	FREIDERICA	18	3061	184	2353	4145	4634	5203	4950	4673	5780	3927
9.	MECKLINGSBANKER E.	10	2851	0	0	0	2622	1257	2635	2913	2280	3422
10.	FRIEDE	8	2682	0	0	0	0	0	1407	3095	3358	2006
11.	V. ENGELSBURG	14	2607	0	585	1899	2234	1985	2385	2715	3632	3605
12.	V. NACHTIGAL	18	2602	477	3389	4216	5201	4481	3923	4443	4400	2145
13.	V. NACHT.&AUFGOTTGEWAGT T.	8	2305	0	0	0	0	0	42	231	292	4319
14.	TRAPPE	18	2256	1596	2028	193	1971	2102	2219	2462	2357	3043
15.	V. TIEFBAU MORGENSTERN O.	11	2151	0	0	0	0	852	1094	2183	3075	1227
16.	JOHANN FRIEDERICH	18	2105	1513	2265	2372	2366	258	2059	1925	2196	2447
17.	FRIEDRICH WILHELM	18	2060	1213	1475	1847	1684	2147	2077	1634	1987	3565
18.	V. CHARLOTTE	13	2047	0	239	3224	3617	3474	913	2568	2549	0
19.	V. BECKEFELD	17	1899	0	610	1291	1786	3024	2759	2416	3189	0
20.	LOUISE TIEFBAU	5	1876	0	0	0	0	0	0	0	0	1876
21.	EINTRACHT	18	1835	133	2312	3569	4878	4502	4045	3917	2828	2067
22.	HASENWINKEL & HIMMEL E.	3	1730	0	0	0	0	0	0	0	0	1730
23.	CLARENBERG	18	1704	143	1332	1557	1423	1895	1559	2656	2957	2386
24.	ELEONORA	17	1694	0	293	142	151	1201	1331	1986	2467	2322
25.	HAMBURG AMTS HORDE	18	1572	1115	1297	1477	1439	1408	1513	1934	2104	1772
26.	V. ANKUNFT & ANKLANG	18	1551	473	5077	669	657	618	643	607	488	166

NAME	YEARS OPERATING 1830–47	AVER. AGE	1830	1836	1837	1838	1839	1840	1841	1842	1847
27. HIMMELSFURSTEN E.	18	1543	1224	2556	3186	2973	2098	2130	2092	149	2245
28. FUNDLING & DAHLACKER	12	1489	0	1613	1818	2136	1800	1896	1679	1366	0
29. SCHURBANK & CHARLOTTENB.	18	1485	1476	1021	397	1589	1739	1230	1494	259	3187
30. KIRSCHBAUM	13	1419	191	0	0	125	1222	1566	2120	2591	1477
31. URBANUS	18	1348	955	1542	1036	1210	1175	680	757	1569	43
32. LOUISENGLUCK	18	1335	785	432	569	1921	1560	1673	1812	1562	2561
33. FRISCHAUF	3	1322	0	0	0	0	0	0	0	0	1322
34. LOUISE	18	1290	770	598	485	904	576	655	1005	1870	1923
35. LUCIA	17	1214	0	516	143	48	49	1455	1673	1341	1786
36. STRALSUND	18	1190	690	220	230	230	148	31	47	50	3800
37. V. NEUE MISSGUNST	18	1183	0	1047	1580	1971	1739	1325	1502	1319	0
38. MORGENSTERN-OSTEN	12	1176	1383	969	444	712	652	0	0	0	0
39. V. PREUSSISCHER ADLER	13	1144	0	1539	2073	2246	1953	1997	1544	1097	796
40. CARL FRIEDRICH E.	16	1088	177	1436	194	1862	1735	1285	1359	1631	1111
41. GLUCKSANGANG III	12	1015	1091	939	748	566	495	154	45	0	0
42. STOCK & SCHERENBERGER H.	18	988	89	1193	137	1079	985	1195	1187	1199	1473
43. WILHELMINE	2	978	0	0	0	0	0	0	0	0	978
44. V. ALTESACKBERG & GEITLING	18	942	759	957	1333	1131	848	869	589	768	1287
45. WOHLVERWAHRT	17	926	170	748	1294	1245	1225	1221	1568	1637	1150
46. FORELLE & ALTEWEIB	18	918	804	480	220	430	813	992	1050	1029	1361
47. EINIGKEIT S.F.	6	893	0	0	0	0	0	0	0	130	1656
48. STORCKSBANK	18	891	301	691	888	681	751	1066	1439	1582	992
49. GULDENE SONNE	11	886	0	433	823	989	956	1112	1423	1339	0
50. GLUCKSBURG	18	886	299	859	972	1304	917	535	1339	1593	796
51. HELLENBANK	15	877	1232	624	784	720	529	561	561	777	0
52. CARL FRIEDERICH E.	10	856	856	0	0	0	20	0	0	0	0
53. SIEPER & MUHLER GRUBE	18	852	832	994	1048	1012	970	783	589	939	645
54. MUHLHEIMERGLUCK	18	834	1455	742	1855	2234	1152	1160	1236	878	262
55. CHRISTIANE	7	830	0	0	0	0	0	0	300	852	809
56. FORTUNA WESTEN	18	813	434	635	573	694	873	738	917	920	1266
57. AUFGOTTGEWAGT	18	805	172	723	601	710	962	1257	1423	1538	788

	NAME	YEARS OPERATING 1830-47	AVER. AGE	1830	1836	1837	1838	1839	1840	1841	1842	1847
58.	WULFINGSBURG	18	751	0	427	384	573	464	649	687	844	982
59.	STOCKER HAUPTGRUBE	18	746	771	1071	943	999	1035	988	983	646	498
60.	V. RUHRMANNSBANK	18	730	866	883	1004	1283	1103	815	1012	570	603
61.	DACHS & GREVELSLOCH	18	717	347	365	293	259	808	1013	1017	1008	1149
62.	GLUCKAUF & SEGEN	13	709	0	243	1729	2160	1446	1125	973	1175	0
63.	KURZES ENDE	4	697	0	0	0	0	0	0	0	0	697
64.	CONCORDIA	18	657	905	343	345	261	712	571	812	877	505
65.	V. SCHELLE & HABERBANK	17	654	795	1098	1113	1025	830	914	478	69	0
66.	GLUCKAUF	10	648	0	0	0	0	0	281	711	683	694
67.	V. ST. GEORG	18	628	523	0	0	0	0	0	0	0	734
68.	ST. PETER	18	599	901	318	117	704	657	577	542	450	728
69.	ROSA	7	581	0	0	0	0	0	0	241	429	734
70.	FORTUNA	7	578	0	0	0	0	0	16	131	578	0
71.	FRIELINGHAUS	18	576	61	1601	2192	2530	1374	602	418	334	308
72.	CRONENBANK	11	563	0	0	197	470	447	592	577	572	555
73.	TULIPAN	18	552	478	578	545	575	477	516	365	502	652
74.	AUGUSTUS	18	544	437	543	529	726	636	500	752	742	457
75.	WIEDERLAGE	12	537	0	354	733	868	658	861	80	720	0
76.	V. ANNA CATHARINA & DICKEB.	16	536	1012	61	0	0	0	0	0	0	0
77.	IDUNA	6	518	0	0	0	0	0	0	0	0	518
78.	CARL WILHELM	5	513	0	0	0	0	0	0	0	0	513
79.	V. GLUCKAUF	18	509	93	820	638	1074	812	842	860	715	409
80.	GEITLING	18	503	1055	325	329	403	365	320	374	343	291
81.	SCHWARTZE JUNGE	9	498	518	479	114	0	0	0	0	0	0
82.	HARMONIE	10	498	0	0	0	29	279	112	125	168	828
83.	DANNENBAUM	18	494	683	292	466	845	1029	1025	866	973	30
84.	JULIANE	9	478	0	0	0	0	0	22	239	314	642
85.	HAGEN SIEPERBANK	15	467	0	159	1288	1106	559	242	796	938	306
86.	EINIGKEIT	14	466	0	0	0	0	0	158	687	913	19
87.	WIENDAHLSNEBENBANK	18	454	68	0	0	0	0	0	4	342	954
88.	GIDEON	14	450	0	526	648	473	545	471	153	151	674
89.	GIBRALTER E.	17	440	0	392	830	1897	506	443	912	920	10
90.	STUCKTEY	17	437	0	429	638	508	515	469	434	487	396

	NAME	YEARS OPERATING 1830-47	AVER. AGE	1830	1836	1837	1838	1839	1840	1841	1842	1847
91.	CHRISTINE&HILFES GOTTES S.F.	3	435	0	0	0	0	0	0	0	435	0
92.	ALTE BOMMERBANK	18	419	675	535	611	648	578	524	276	227	242
93.	DIEDERICH	10	416	0	0	0	0	0	0	315	418	414
94.	REIGER	11	409	302	0	0	0	0	0	0	0	516
95.	PIEPERS E.	6	406	0	0	0	0	0	0	0	33	779
96.	ALTE SACKBERG	18	393	584	617	980	1033	730	354	337	308	66
97.	NACHTIGALL	18	391	509	565	810	530	523	375	270	399	92
98.	STEINBERGERBANK	16	390	328	498	520	473	278	37	0	0	345
99.	FRANZISKA FORTSETZUNG	11	381	0	0	393	872	676	700	584	630	133
100.	ALTE APROCHE	18	380	436	397	18	394	978	698	990	278	409
101.	WESELBANK	1	364	364	0	0	0	0	0	0	0	0
102.	ALTE HAASE	8	361	296	427	177	0	0	0	0	0	0
103.	ST. MORTIZ	18	354	262	298	254	348	264	288	275	520	338
104.	ALTEMANN	11	351	0	425	849	983	811	531	591	277	0
105.	V. HIMMELFAHRT	4	346	346	0	0	0	0	0	0	0	0
106.	NEUGLUCK & STETTIN	18	340	272	389	290	291	362	489	410	398	302
107.	STOCKERDRECKBANK	10	333	473	193	205	188	88	0	0	0	0
108.	V. ALEXANDER	8	325	0	325	266	228	269	190	103	0	0
109.	HEINRICH	7	318	0	0	0	0	0	0	61	172	465
110.	N. SACKBERG & SCHRUTENBERG	9	316	0	0	0	0	355	654	534	450	183
111.	V. WILDENBERG & VOGELBRUCH	15	314	315	0	2	31	0	0	153	314	0
112.	BUNTEBANK	16	311	113	0	11	180	398	214	376	509	0
113.	TURTELTAUBE N.	9	310	310	0	16	0	0	0	0	0	0
114.	TURTELTAUBE S.	6	305	305	0	0	0	0	0	0	0	0
115.	FRANZISKA	18	304	474	437	150	222	270	230	0	2	0
116.	GLUCKSONNE	9	302	265	339	501	175	0	0	0	0	0
117.	TIMMERBEIL	18	297	225	99	125	109	137	156	86	81	785
118.	JUPITER	3	293	0	0	0	0	0	0	0	0	293
119.	NEUGLUCK	18	291	436	259	224	312	245	215	200	168	303
120.	GEDULD	7	284	0	0	1	14	46	12	54	284	0

	NAME	YEARS OPERATING 1830-47	AVER. AGE	1830	1836	1837	1838	1839	1840	1841	1842	1847
121.	HERMANNS GES. SCHIFFAHRT	18	283	119	120	130	159	159	168	182	227	666
122.	NEUE STEINKUHL	16	282	0	555	560	182	282	8	205	74	16
123.	ISABELLE	7	281	0	9	12	0	0	0	0	0	553
124.	HAZARD	13	269	178	249	348	318	243	262	438	380	0
125.	ST. MATHIAS	18	263	29	194	621	449	714	437	655	819	13
126.	SCHWARZER ADLER	17	259	156	228	347	329	267	266	316	393	0
127.	NEUE GOTTSEGENEDICH	18	259	160	164	92	23	134	325	522	453	0
128.	KONIGSCRONE	4	253	0	253	491	37	0	0	0	0	0
129.	VOGELBRUCH	14	253	0	246	199	302	240	271	112	0	261
130.	NIEDERHOFEN	18	249	295	336	42	376	480	387	283	265	102
131.	ALTE APROCHE SCHACHT	1	246	246	0	0	0	0	0	0	0	0
132.	KAISBERGERBANK	3	246	0	246	157	150	0	0	0	0	0
133.	AUFGOTTGEWAGT TIEFBAU	2	243	0	243	632	0	0	0	0	0	0
134.	JOHANNES E.	8	242	307	178	65	0	0	0	0	0	0
135.	LEVERINGSBANK	9	228	401	56	222	33	61	0	0	0	0
136.	V. SCHIERBANK	18	226	108	282	244	59	121	219	231	268	247
137.	BERGMANN	5	223	0	0	0	0	0	0	0	0	223
138.	ELIAS E.	17	220	0	194	26	0	1	132	500	298	168
139.	V. FIEDRICHSFELD	18	220	0	0	3	0	5	73	151	202	239
140.	HAMMERBANK	14	216	0	143	260	386	464	430	470	289	0
141.	GLUCKAUF TIEFBAU	7	215	0	0	0	0	0	0	118	172	259
142.	FELICITAS	7	208	364	52	0	52	0	0	0	0	0
143.	DANNENBAUM N.F.	1	208	0	0	0	0	0	0	0	0	208
144.	ELISABETH	8	208	386	31	0	9	0	0	0	0	0
145.	BUSCHBANK	18	203	105	163	221	217	366	280	295	245	300
146.	TREUE	18	202	79	196	348	309	267	347	489	506	27
147.	ST. JOHANNES E.&FRIELING.	3	196	196	0	0	0	0	0	0	0	0
148.	VERLORENE POSTEN	10	193	193	0	8	72	33	0	0	0	0
149.	SCHWARZE PFERD	4	189	189	0	0	0	0	0	0	0	0
150.	HESSENBANK	3	186	186	0	0	0	0	0	0	0	0

	NAME	YEARS OPERATING 1830-47	AVER. AGE	1830	1836	1837	1838	1839	1840	1841	1842	1847
151.	PORTBANKER E.	18	184	184	0	0	0	0	0	0	0	0
152.	NACHGEDACHT	1	183	0	0	0	0	0	0	0	0	0
153.	CHRISTIANE&HILFE GOTTES	15	181	0	168	1908	2128	1919	1054	883	195	0
154.	KIRSBAUM	4	178	178	0	0	0	0	0	0	0	0
155.	RABENER	7	174	0	0	0	0	29	218	129	174	0
156.	ST. PAUL	5	173	0	0	0	0	0	0	0	0	173
157.	ST. JOHANNES E.	18	171	51	97	706	569	595	352	318	273	265
158.	FROSCH	4	162	162	0	0	0	0	0	0	0	0
159.	CHRISTINE I	4	160	160	0	0	0	0	0	0	0	0
160.	HEINRICH WILHELM	11	158	0	190	235	201	177	103	70	127	0
161.	ST. JOSEPH	18	157	120	316	259	171	191	195	183	115	77
162.	DESTERBANK	7	156	150	162	35	0	0	0	0	0	0
163.	CHARLOTTE	13	154	131	286	401	335	422	161	183	47	0
164.	HAMBURG (WITTEN)	7	151	0	151	571	654	255	0	0	0	0
165.	SIEBENPLANETEN	16	150	150	0	0	0	0	55	148	0	0
166.	HEINRICH S.F.	2	145	0	0	0	0	0	0	0	0	145
167.	ST. GEORG	18	143	66	0	103	122	138	155	175	21	0
168.	ALTE BOMMERBANK E.	18	141	0	141	174	93	188	216	27	0	0
169.	GLUCKSANFANG II	12	124	171	78	130	167	26	84	91	0	0
170.	FEIGENBAUM	7	118	0	0	0	0	63	70	210	118	0
171.	FLACHSTEICH	4	115	115	0	0	0	0	0	0	0	0
172.	RUMMELSKIRCHEN	9	115	45	186	75	7	0	0	0	0	0
173.	CAROLINE	18	113	280	124	96	53	0	4	25	32	16
174.	LOWE	18	110	20	222	187	184	217	239	212	159	40
175.	EGMONTS E.	3	107	0	0	0	0	0	0	0	0	107
176.	FRIEDERICH	17	105	0	149	319	500	217	173	268	137	29
177.	V. DICKEBACKERBANK	2	103	0	103	247	0	0	0	0	0	0
178.	ANGST & BANGE	9	101	0	0	0	0	4	159	128	145	57
179.	SCHLIGGE	4	100	125	75	32	0	0	0	0	0	0
180.	FROHE ANSICHT	18	98	38	136	99	184	125	188	109	175	44

	NAME	YEARS OPERATING 1830-47	AVER. AGE	1830	1836	1837	1838	1839	1840	1841	1842	1847
181.	IGNATIUS	10	95	0	95	354	219	122	45	0	0	0
182.	DRECKBANKER E.	7	92	0	0	0	0	0	0	0	0	92
183.	ELEONORA & NACHTIGALL	4	91	0	0	0	0	0	0	0	0	91
184.	HELENA	17	86	86	0	0	0	0	16	48	0	0
185.	BRAUNSCHWEIG N.	4	82	0	0	0	0	0	0	0	0	82
186.	CLAUSTHAL	7	81	0	160	326	335	427	351	175	2	0
187.	JULIUS PHILIP E.	9	74	0	0	0	0	0	0	0	124	24
188.	EINTRACHT S.F.	2	73	0	0	0	0	0	0	0	73	0
189.	DRECHBANK	3	69	69	0	0	0	0	0	0	0	0
190.	CHRISTIAN	4	66	66	0	0	0	0	0	0	0	0
191.	BUSCHBANK NEBENBANK	4	65	0	0	0	0	0	0	0	0	65
192.	OBERLEVERINGSBANK	1	58	0	0	0	0	0	0	0	0	58
193.	KLEINGLUCK	3	55	0	0	0	0	0	22	81	55	0
194.	V. FRIEDRICH WILHELM	12	55	0	104	83	25	1	0	1	7	0
195.	MARGARETHE	14	53	116	26	18	9	0	0	2	17	0
196.	SENGSTBANK	2	36	36	0	0	0	0	0	0	0	0
197.	MAXIMUS	11	32	0	0	0	0	115	208	243	32	0
198.	VERLOHRNER SOHN	4	31	0	0	0	0	0	47	80	31	0
199.	ADOLPHUS	3	30	30	0	0	0	0	0	0	0	0
200.	V. GROSSE VARST. & ALTEN E.	18	28	28	0	1	1	3	0	0	0	0
201.	CRONE	1	27	0	0	0	0	0	0	0	0	27
202.	SIEBEN SOHNE	1	23	0	0	0	0	0	0	0	23	0
203.	SANDBERG	2	17	0	0	0	0	0	0	0	4	31
204.	HAMBURG S.F.	1	16	16	0	0	0	0	0	0	0	0
205.	KURZEGGERBANK	3	16	0	16	37	159	0	0	0	0	0
206.	PAPENBANK	4	16	0	0	0	0	0	0	0	0	16
207.	BELLE ALLIANCE	1	15	0	0	0	0	0	0	0	0	15
208.	GLUCKAUFER E.	13	11	0	0	0	0	0	24	19	18	4
209.	V. EMMANUEL	2	9	0	0	0	0	0	0	0	0	9
210.	MARTHA	2	8	0	0	0	0	0	0	0	0	8
211.	SCHIFFSRUDER	5	7	0	1	5	0	0	0	0	0	13
212.	BLUMENTHAL	1	5	0	0	0	0	0	0	0	0	5
213.	HELLER MITTAG	3	4	0	0	0	0	0	0	0	0	4

	NAME	YEARS OPERATING 1830-47	AVER. AGE	1830	1836	1837	1838	1839	1840	1841	1842	1847
214.	CATHARINE	12	4	0	8	0	0	2	5	0	1	0
215.	V. LOUISE	4	3	0	0	0	0	0	0	0	0	3
216.	HALTER	11	1	0	0	0	0	0	0	0	0	1
217.	AUFGOTTGEWAGT S.F.	3	0	0	0	0	0	1645	0	0	0	0
218.	MORGENSTERN WESTEN	3	0	0	0	11	835	810	0	0	0	0
219.	ST. JOHANNES E. S.F.	3	0	0	0	331	842	0	0	0	0	0
220.	VEREINIGUNGS STOLLEN	2	0	0	0	0	0	0	0	0	0	0
221.	FRIEDRICH W. FETTE HENNE	1	0	0	0	0	0	0	0	0	0	0
222.	CARTHAUSERLOCH	1	0	0	0	0	0	0	0	0	0	0
223.	SONNENSCHEIN	2	0	0	0	0	0	0	0	0	0	0
224.	CLEVERBANK	1	0	0	0	0	0	0	0	0	0	0
225.	ALTEMANN S.F.	1	0	0	0	0	0	0	0	0	0	0
226.	STRAUSS	1	0	0	0	0	0	0	0	0	0	0
227.	FRIEDLICHER NACHBAR	1	0	0	0	0	0	0	0	0	0	0
228.	JOHANN MATHIAS	1	0	0	0	0	0	0	0	0	0	0
229.	PREUSSISCHER ZEPTER	1	0	0	0	0	0	0	0	0	0	0
230.	VOGELHERDT	1	0	0	0	0	0	0	0	0	0	0
231.	V. CAROLINE & ADLER	1	0	0	0	0	0	0	0	0	0	0
232.	KRACKHAUSBANK	1	0	0	0	0	0	0	0	0	0	0
233.	FRIEDERICH	2	0	0	0	0	0	0	0	0	0	0
234.	GUTE KAUF	2	0	0	0	12	112	0	0	0	0	0
235.	WASSER & WASSERNEBENBANK	2	0	0	0	0	0	0	0	0	0	0
236.	JUNGE WITWE	2	0	0	0	0	0	0	0	0	0	0
237.	SCHULTENKAMPERBANK	2	0	0	0	0	0	0	0	0	0	0
238.	GLUCKSFORTGANG	2	0	0	0	0	0	32	87	0	0	0
239.	FERNER GLUCK	2	0	0	0	0	0	5	7	0	0	0
240.	LUSTIG	1	0	0	0	0	0	0	0	0	0	0
241.	AM SCHWABEN	2	0	0	0	0	0	0	0	0	0	0
242.	CARL WILHELM E.	3	0	0	0	0	0	0	0	0	0	0
243.	GLUCKAUFS E. FLUGELORT	2	0	0	0	0	0	0	0	0	0	0
244.	NICOLAUS & VICTORIA	2	0	0	0	0	0	0	0	0	0	0
245.	V. KONIGSCRONE	3	0	0	0	0	7	37	124	0	0	0

NAME	YEARS OPERATING 1830-47	AVER. AGE	1830	1836	1837	1838	1839	1840	1841	1842	1847
246. TIEFBAU MULDENPARTIE	2	0	0	0	0	0	0	0	0	0	0
247. WECKLENBANK	3	0	0	0	0	0	0	0	0	0	0
248. HORSTER E.	12	0	0	0	0	0	0	0	0	0	0
249. ST. JOHANNES E.&FRIEL. T.	16	0	0	0	0	0	535	0	0	0	0

Source: Monthly Mining Reports, 1830-47

TABLE

1836-42, ANNUAL OUTPUT FOR ESSEN COAL OPERATIONS (100 Scheffel)

(Arranged by average output)

	AVERAGE	1836	1837	1838	1839	1840	1841	1842
1. GEWALT	5400	4624	5200	5504	5600	6200	3460	7252
2. SAELZER - NEUAK	5328	5172	4216	4872	5904	4944	5576	6604
3. SCHOLERPAD	4504	1920	2864	3880	5688	6652	4696	5820
4. KUNSTWERK	4360	3488	1792	5376	5492	4964	5044	4304
5. V. HENRIETTE	3692	3228	600	6444	4844	4696	1708	4328
6. LANGENBRAHM	3620	3688	2876	3652	3328	3076	4032	4692
7. PORTINGSIEPEN	2960	2560	2748	2944	2908	2836	3448	3224
8. GRAF BEUST	2532	0	0	0	0	0	0	2532
9. HITZBERG & DICKEB.	2412	2400	2028	2004	1936	4524	2004	1976
10. HUNDSKNOCKEN	2260	1860	480	2100	1440	2468	3716	3760
11. STEINGATT UNTER MARK	1796	1444	580	1300	924	2768	2644	2920
12. STEINGATT	1552	600	2220	3200	3144	904	412	1180
13. ILANDSBRAUT	1528	800	1300	1040	120	1896	2160	2344
14. SCHWARZE ADLER	1228	624	864	1496	120	1696	1536	1172
15. WOHLGEMUT	1192	1192	0	0	0	0	0	0
16. V. GIRONDELLE	1140	1084	260	1580	1444	1124	1456	940
17. BRUCHKAMP & STEINKNAPPE	1128	1364	1260	1304	1324	1016	488	0
18. V. HERMANN	1056	0	0	408	736	1464	1680	988
19. PREUTEN BORBECK SIEPEN	960	228	480	804	1152	1588	1800	1860
20. KAMPGESBANK	892	384	500	884	532	1140	1292	1516
21. BALDENEIER E.	880	312	120	880	960	1084	1480	1244
22. V. HOBEISEN	880	296	240	848	1252	844	1808	0
23. WOLFSBANK	828	0	0	0	0	20	916	1552
24. KRUSENBERG	808	0	0	0	0	0	320	1300
25. AMSEL & SCHNEPPE	736	0	0	0	0	0	8	736

	AVERAGE	1836	1837	1838	1839	1840	1841	1842
26. V. CAPPELLENBANK	732	664	696	900	856	708	736	548
27. DUVENKAMPSBANK	732	400	128	896	904	852	1028	908
28. NIERBANK	672	0	4	280	576	724	1772	672
29. BRAUT - KUPERSWIESE	572	520	488	504	568	512	488	924
30. MAASBANK	564	0	0	0	72	400	920	880
31. V. HIPPE	560	756	716	668	332	576	280	0
32. FLOTE	512	480	0	24	0	0	0	0
33. STEINGATT UNTER HEINRICHSFELD	488	0	0	0	244	976	252	0
34. V. LOUISE	468	0	0	0	40	1136	156	108
35. PLATZGESBANK NEBENFLOTZ	448	252	0	104	408	592	556	776
36. SAELZER - NEUAK S.F.	448	0	0	0	456	924	192	212
37. NEUGLUCK	424	376	404	512	480	212	212	772
38. KRONENPRINZ V. PREUSSEN	416	0	0	280	216	596	908	72
39. PFAFFENBERG	400	0	16	516	588	536	304	0
40. NEUGLUCK S.F.	360	0	0	0	0	360	0	0
41. KALSKSIEPEN	356	16	124	120	212	576	720	724
42. VEREINIGUNG UBER STOLLENSOHLE	344	0	20	144	0	0	0	0
43. PLATZGESBANK H.	328	328	0	0	0	0	0	0
44. TROMPETE	312	280	484	392	196	0	36	480
45. ANSELMUS E.	308	0	48	304	352	440	412	288
46. TROMPETE UNTER MARK	296	180	136	276	336	420	436	0
47. BRAUT & NEBENBANK	280	564	420	276	448	208	0	0
48. NIPSHAGEN	280	280	0	0	0	0	0	0
49. V. KUCKS & SCHINKENBANK	268	324	420	304	260	168	136	0
50. BETTELSTRUCKER ST.	256	100	72	132	308	448	276	448
51. BEISENKAMP	232	188	52	300	220	364	320	180
52. ERENBANK	228	212	232	240	228	292	252	132
53. SCHUTZDELLERBANK	228	156	392	0	88	140	356	0
54. NOTTEKAPPE	212	0	800	336	312	296	44	280
55. SCHINKENBANK	208	92	264	264	396	192	40	0
56. FORTUNA	192	184	0	8	0	0	0	0
57. HEIMANNSFELD	192	0	0	4	80	24	444	404
58. V. HOFFNUNG	184	472	28	360	176	104	20	120
59. ERBENKAMPSBANK	176	176	0	0	0	0	0	0
60. JACKE	176	160	236	264	240	56	136	136

	AVERAGE	1836	1837	1838	1839	1840	1841	1842
61. GROSSE BOVERMANNSWIESE	164	0	0	0	0	24	308	164
62. SCHMACHTENBANK	160	144	140	208	164	120	148	164
63. V. HAGENBECK	156	0	0	0	0	96	308	160
64. STEINGATT DICKEB. SUDFL.	156	240	132	0	180	68	0	0
65. TROTZ	136	188	344	76	4	76	0	0
66. MUNLE	128	128	0	0	0	0	0	0
67. V. DEIMELSBERGER E.	116	132	20	264	200	40	20	136
68. HERRENBANK	108	108	100	96	108	76	68	196
69. CATHARINA LEVEMANNSFELD	96	0	0	0	0	0	112	80
70. V. FELD	88	88	0	0	0	0	0	0
71. VEREINIGUNG UNTER STOLLENSOHLE	80	0	16	12	0	0	52	1648
72. KAHRERWIEDERTOPFE	68	68	0	0	0	0	0	0
73. FRISCHES GLUCK & FRIEDERICH	60	60	76	80	32	0	0	0
74. PAULINA E.	60	0	0	56	80	96	68	0
75. WILHELMINE SCHNEPPE	52	0	0	0	0	52	0	0
76. V. CATHARINA	36	4	0	16	16	4	4	200
77. PAUL	24	0	0	0	0	20	16	36
78. MONCHHOFFSBANK	20	0	0	0	36	0	0	4
79. UBEREINKUNFT	16	0	16	0	0	0	0	0
80. V. PAUL & V. HIPPE ST.	12	4	4	24	32	4	0	0
81. PRINZ FRIEDRICH	12	0	0	0	0	0	0	12

Source: Annual Mining Reports, 1823-43

TABLE

1836-42 AVERAGES FOR ENTIRE RUHR COAL OPERATIONS

(Arranged by Profit Order)

	NAME	YEARS OF OPERATION	OUTPUT	INCOME	COST	PROFIT	TAXES	TECHNOL.	DISTRICT
1.	V. HASENWINKEL & HIMMEL	(0 0) 18	578776	78466	49378	29088	8673	S R I	7
2.	SAELZER - NEUAK	(0 0) 7	532800	72511	44551	27960	9976	S R I	17
3.	V. NACHTIGAL	(0 0) 18	429384	50574	28695	21879	6764	S R I	10
4.	GLUCKAUF	(0 0) 18	333672	41876	22003	19873	5982	S I	3
5.	V. GENERAL & ERBST.	(0 0) 18	555976	62041	43604	18437	7558	R I	7
6.	FRIEDERICA	(0 0) 18	443584	48700	32885	15815	7056	S I	5
7.	TRAPPE	(0 0) 18	215280	34266	20118	14148	4297	S I	13
8.	SCHOLERPAD	(0 0) 7	450400	57542	43819	13723	7994	S R I	17
9.	EINIRACHT	(0 0) 18	372236	42237	28748	13489	5386	R I	8
10.	FRIEDE	(40 0) 8	221988	28420	16439	11981	3902	R I C	10
11.	KUNSTWERK	(0 0) 7	436000	52759	42088	10671	7286	S R I	17
12.	JOHANN FRIEDERICH	(0 0) 18	224156	28520	19508	9112	3133	R I	7
13.	STOCK & SCHERENBERGER H.	(0 0) 18	117244	22233	13752	8481	2817	I C	12
14.	HITZBERG & DICKEB.	(0 0) 7	241200	17767	9477	8290	1014	R I	18
15.	STEINGATT UNTER MARK	(0 0) 7	179600	15525	7577	7948	2074	S R I	16
16.	V. TIEFBAU MORGENSTERN	(37 0) 11	180104	19444	11611	7833	2733	R I C	10
17.	GEWALT	(0 0) 7	540000	66652	59694	6958	7901	S R I	16
18.	LANGENBRAHM	(0 0) 7	362000	31881	24977	6904	4627	S R I	18
19.	STOCKER HAUPIGRUBE	(0 0) 18	80420	17913	11217	6696	2325	I C	12
20.	MECKLINGSBANKER E.	(38 0) 10	227124	24396	18071	6325	3172	R I C	8
21.	FUNDLING & DAHLACKER	(32 43) 12	152675	18056	12400	5656	2575	S C	2
22.	GLUCKAUF & SEGEN	(35 0) 13	126436	15679	10111	5568	2388	S I C	3
23.	HAMBURG AMTS HORDE	(0 0) 18	159648	15249	9737	5512	2302	R	4
24.	FRIELINGHAUS	(0 0) 18	130712	14921	9524	5397	1909	S R	10

	NAME	YEARS OF OPERATION	OUTPUT	INCOME	COST	PROFIT	TAXES	TECHNOL.	DISTRICT
25.	PORTINGSIEPEN	(0 0) 7	296000	23328	18041	5287	3560	S R I	14
26.	KIRSCHBAJM	(0 0) 13	152480	18886	14083	4803	2298	R I C	7
27.	HUNDSNACKEN	(0 0) 7	226000	20323	15551	4772	3146	S R I	15
28.	V. SCHELLE & HABERBANK	(0 45) 17	79496	12036	7398	4638	1770	S	11
29.	HIMMELSFURSTEN E.	(0 0) 18	236216	20567	16272	4395	3026	R I	9
30.	SIEPER & MUHLER GRUBE	(0 0) 18	90515	18040	13748	4292	2273	R I	12
31.	AUFGOTTGEWAGT	(0 0) 18	103136	11318	7574	4244	1635	R I	10
32.	CLARENBERG	(0 0) 18	191132	22252	18159	4093	3258	S	3
33.	V. HEUE MISSGUNST	(0 0) 18	136280	15930	11845	4085	1844	R I	6
34.	AUFGOTTGEWAGT S.F.	(38 40) 3	164548	15976	11904	4072	2176	R	10
35.	MUHLHEIMERGLUCK	(0 0) 18	132348	13907	10245	3662	1918	R I	9
36.	ST. PETER	(0 0) 18	48080	9593	6055	3638	1320	S I	13
37.	CHRISTIANE & HILFE GUTTES	(32 46) 15	117916	16819	13205	3614	1991	R I C	7
38.	V. ALTESACKBERG & GEITLING	(0 0) 18	92780	10456	7178	3278	1349	R I C	9
39.	WOHLVERWAHRT	(31 0) 17	127684	12535	9257	3278	1810	R I C	8
40.	GULDENE SONNE	(33 45) 11	101076	11065	7865	3200	1673	I C	5
41.	MORGENSTERN-OSTEN	(0 40) 12	69404	7149	4058	3091	1021	R I	10
42.	V. RUHRMANNSBANK	(0 0) 18	95280	9043	5999	3044	1295	R	4
43.	NACHTIGALL	(0 0) 18	49092	9543	6932	2611	1237		11
44.	STORCKSBANK	(30 0) 18	101416	11265	8776	2489	1609	I C	3
45.	V. PREUSSISCHER ADLER	(34 0) 13	177836	15924	13448	2476	1883	R I C	9
46.	SAELZER - NEUAK S.F.	(39 0) 4	44800	6872	4442	2430	937	I	17
47.	N. SACKBERG & SCHRUTENG.	(39 0) 9	49824	6352	3932	2420	853	R	9
48.	FORTUNA WESTEN	(0 0) 18	76428	7351	4947	2404	1116	R I	10
49.	ELEONORA	(31 0) 17	108176	11947	9556	2391	1711	S C	2
50.	CHRISTIANE	(41 0) 7	57588	5723	3426	2297	853		4
51.	V. BICKEFELD	(31 0) 17	215368	23563	21357	2206	3471	S C	2
52.	BRUCHKAMP & STEINKNAPPE	(0 41) 6	112800	6987	4813	2174	1224	R I	15
53.	CARL FRIEDERICH E.	(0 41) 10	135736	16982	14890	2092	1756	R I	6
54.	CARL FRIEDRICH E.	(32 0) 16	135736	16982	14890	2092	1757	R I C	6
55.	V. ANKUNFT & ANKLANG	(0 0) 18	125128	6959	4908	2051	993	R	10
56.	GLUCKSBURG	(0 0) 18	107404	10750	8791	1959	1282	R I C	6
57.	MORGENSTERN WESTEN	(37 39) 3	80956	8860	6906	1954	1239	C	10

	NAME	YEARS OF OPERATION	OUTPUT	INCOME	COST	PROFIT	TAXES	TECHNOL.	DISTRICT
58.	V. GROSSE VARST. & ALTEN	(0 0) 18	188	3699	1765	1934	3	R I	9
59.	BRAUT - KUPERSWIESE	(0 0) 7	57200	5136	3292	1844	780	R	14
60.	ST. JOHANNES E. S.F.	(37 39) 3	20744	5150	4320	1830	838	R	10
61.	V. GLUCKAUF	(0 0) 18	70036	7495	5689	1806	1147		11
62.	GIOEON	(34 0) 14	42360	4585	2813	1772	645		11
63.	GLUCKSONNE	(0 38) 9	33860	3994	2222	1772	562	R	7
64.	ALTE APROCHE	(0 0) 18	68924	5069	4374	1695	823	R	9
65.	LOUISE	(0 0) 18	87060	10350	8573	1677	1440	I	1
66.	TURTELTAUBE N.	(30 38) 9	29724	3438	1776	1662	481	R C	10
67.	BALDEMEIER E.	(0 0) 7	88000	5512	4857	1655	1095	R I	15
68.	STUCKTEY	(31 0) 17	49704	4528	2907	1621	694	I C	4
69.	ALTE SACKBERG	(0 0) 18	62284	6569	4952	1617	762	R	9
70.	BUNTEBANK	(0 45) 16	28128	4072	2456	1616	554	I	3
71.	SCHWARZE ADLER	(0 0) 7	122800	5930	5381	1549	1157	R I	16
72.	V. GIRONDELLE	(0 0) 7	114000	8306	6761	1545	1333	R	15
73.	STEINGATTE UNTER H.	(39 41) 3	48800	4782	3264	1518	727	R I	16
74.	BRAUT & NEBENBANK	(0 40) 5	28000	3413	1960	1453	559	R	14
75.	DACHS & GREVELSLOCH	(0 0) 18	68052	10490	9049	1441	1423	S I	13
76.	FRANZISKA	(0 0) 18	26176	3870	2447	1423	328	R	4
77.	FORELLE & ALTEWEIB	(0 0) 18	71624	8107	6591	1416	1181	S I	1
78.	RABENER	(39 45) 7	13756	2471	1097	1374	182	R	10
79.	ILANDSBRAUT	(0 0) 7	152800	11273	9928	1345	892	S R	15
80.	CONCORDIA	(0 0) 18	56008	7556	6253	1803	2154	I C	12
81.	GLUCKAUF	(38 0) 10	53192	10465	9189	1276	1360	I C	12
82.	DANNENBAUM	(0 0) 18	78516	6417	5147	1270	1035	I	5
83.	WIEDERLAGE	(35 0) 12	61100	8158	6895	1263	1212	R I C	10
84.	FRANZISKA FORTSETZUNG	(37 0) 11	64256	5831	4570	1261	872	R C	4
85.	DUVENKAMPSBANK	(0 0) 7	73200	5895	4547	1248	971	R	15
86.	ALTE BOMMERBANK	(0 0) 18	48552	4073	2871	1202	648		13
87.	WIENDAHLSNEBENBANK	(0 0) 18	17336	1846	661	1175	241	I	3
88.	GLUCKSANGANG III	(0 41) 12	49104	4931	3805	1126	727		5
89.	NEUGLUCK S.F.	(0 40) 1	36000	2210	1086	1124	404	R I	16
90.	ST. JOHANNES E.	(0 0) 18	52564	8746	7637	1109	813	S R I	10

	NAME	YEARS OR OPERATION	OUTPUT	INCOME	COST	PROFIT	TAXES	TECHNOL.	DISTRICT
91.	WOHLGEMUT	(0 36) 1	119200	4729	3629	1100	611	S I	16
92.	HELLENBANK	(0 44) 15	66376	7136	6067	1069	1034	S	2
93.	KAMPGESBANK	(0 0) 7	89200	6909	5860	1049	1140	R I	18
94.	AUGUSTUS	(0 0) 18	63260	5668	4645	1023	856	R	4
95.	ST. MORTIZ	(0 0) 18	28480	2963	1956	1007	462		3
96.	ALTE HAASE	(0 37) 8	30220	3429	2456	973	501		11
97.	ALTE BOMMERBANK E.	(0 0) 18	48552	4072	3157	915	590	C	13
98.	NEUGLUCK & STETTIN	(0 0) 18	37556	3500	2707	893	549	R	10
99.	NEUGLUCK	(0 0) 18	37556	3600	2707	893	552		12
100.	V. FELD	(35 37) 1	8800	1329	487	842	218	R	15
101.	TULIPAN	(0 0) 17	59735	4149	3395	735	555		11
102.	JOHANNES E.	(0 37) 8	12156	1134	335	799	176	R	4
103.	NIERENBANK	(0 0) 6	67200	3223	2431	792	499	I	14
104.	KONIGSCRONE	(35 38) 4	26056	2537	1844	693	361	C	6
105.	ST. JOHANNES E. & F.T.	(32 0) 16	53488	5994	5324	670	806	S R I C	10
106.	SFAFFENBERG	(37 41) 5	40000	3102	2439	663	540	R I	15
107.	GRONENBANK	(37 0) 11	47572	3983	3347	636	622	C	13
108.	V. ANNA CATHARINA & D.	(0 0) 16	6076	1769	1136	633	156	R I	7
109.	ALTEMANN	(33 43) 11	63824	5623	5016	607	712	R I C	¢
110.	STEINGATT DICKEB. S.	(0 40) 5	15600	1280	689	591	221	R I	16
111.	V. ALEXANDER	(34 41) 8	23032	1798	1227	571	319	R C	4
112.	FLUTE	(0 39) 4	51200	1239	689	550	261	R	18
113.	VOGELBRUCH	(34 0) 14	22840	2190	1659	531	331	C	12
114.	V. KUCKS & SCHINGENB.	(0 0) 6	26800	2126	1600	526	332	R	14
115.	FRIEDRICH WILHELM	(0 0) 18	183592	24223	23716	507	3006	S C	1
116.	HERMANNS GES. S.	(0 0) 18	16360	1804	1304	500	259		11
117.	GIBRALTER E.	(31 0) 17	91884	7282	6798	484	942	R I C	6
118.	CHARLOTTE	(0 42) 13	21084	2210	1780	430	350		11
119.	HAMMERBANK	(31 44) 14	34876	3262	2832	430	460	R I C	6
120.	STRALSUND	(0 0) 18	13088	1564	1161	403	465	R	10
121.	V. HENRIETTE	(0 0) 7	369200	43231	42830	401	6375	S R I	16
122.	CLAUSTHAL	(36 42) 7	29924	3617	3249	368	476	I C	3

	NAME	YEARS OF OPERATION	OUTPUT	INCOME	COST	PROFIT	TAXES	TECHNOL.	DISTRICT
123.	LEVERINGSBANK	(0 38) 9	10560	1320	953	367	188		12
124.	PLATZGESBANK H.	(0 37) 2	32800	1002	645	357	194	R	16
125.	LUCIA	(31 0) 17	74640	8913	8558	355	1071	R I C	7
126.	SCHINKENBANK	(0 41) 6	20800	1706	1356	350	261		14
127.	HAMBURG (WITTEN)	(39 45) 7	40795	4274	3930	344	633	C	4
128.	GLUCKSANFANG II	(0 41) 12	9616	1155	811	344	157		5
129.	NIEDERHOFEN	(0 0) 18	30988	3359	3024	335	627		3
130.	V. DICKEBACKERBANK	(41 42) 2	17516	1533	1203	330	220	R	7
131.	ROSA	(41 0) 7	33500	2507	2305	302	417		13
132.	SCHURBANK CHARL.	(0 0) 18	107492	12430	12161	269	1707	S C	2
133.	ST. JOSEPH	(0 47) 18	20448	1977	1719	258	305		11
134.	FRIEDERICH	(31 0) 17	25216	2454	2228	226	258	C	6
135.	DESTERBANK	(31 37) 7	9828	831	507	224	136		10
136.	ELISABETH	(0 39) 8	1980	498	275	223	93		2
137.	FORTUNA	(0 38) 2	19200	784	577	207	152	R	18
138.	PREUTEN BORBECK	(0 0) 7	96000	6506	6303	203	1002	R I	14
139.	STOCKERDRECKBANK	(0 39) 10	16840	2857	2655	202	378		12
140.	PORTBANKER E.	(0 0) 18	0	548	362	186	0	R	4
141.	BUSCHBANK	(0 0) 18	25516	3300	3115	185	444		12
142.	SCHMACHTENBANK	(0 0) 7	16000	1151	980	171	192	R	18
143.	ERBENBANK	(0 0) 7	22800	1766	1598	168	294	R	18
144.	SCHWARZER ADLER	(30 46) 17	30648	3291	3124	167	218	C	2
145.	HAZARD	(31 43) 13	31976	2904	2738	166	506	R	10
146.	UBEREINKUNFT	(0 37) 2	1600	302	142	160	57	R	16
147.	STEINBERGERBANK	(0 0) 16	36124	2980	2821	159	482		4
148.	ANGST & BANGE	(39 0) 9	14384	1159	1001	158	139	C	3
149.	JACKE	(0 0) 7	17600	1524	1370	154	244	R	14
150.	GEITLING	(0 0) 18	35144	3156	3030	126	494	R	9
151.	RUMMELSKIRCHEN	(0 38) 9	8940	762	649	113	110		11
152.	LOWE	(0 0) 18	20292	2312	2202	110	343		13

	NAME	YEARS OF OPERATION	OUTPUT	INCOME	COST	PROFIT	TAXES	TECHNOL.	DISTRICT
153.	NEUE GOTTSEGENEDICH	(0 0) 18	24460	2100	2003	97	312		11
154.	KAISBERGERBANK	(36 38) 3	19412	1517	1422	95	245	R	10
155.	KAHRERWIEDERTOPFE	(0 41) 6	6800	399	312	87	83	R	14
156.	HERRENBANK	(0 0) 7	10800	542	489	53	98	R	17
157.	PLATZGESBANK BEVENF	(0 0) 7	44800	2522	2471	51	383	R	16
158.	SIEBENPLANETEN	(0 0) 16	10188	309	764	45	131		4
159.	FELICITAS	(0 36) 7	5168	152	132	20	30		2
160.	NEUGLUCK	(0 0) 7	42400	2332	2820	12	475	R I	16

Source: Monthly Mining Reports, 1830-47
Annual Mining Reports, 1823-43

Explanation:
Output in Scheffel; Income- Reichsthaler;
Taxes are included in cost.

TABLE

1836-42, AVERAGES FOR ENTIRE RUHR COAL OPERATIONS

(Arranged by Order of Losses)

	NAME	YEARS OF OPERATION 30-37	OUTPUT	INCOME	COST	LOSSES	TAXES	TECHNOL.	DISTRICT
229.	HIMMERBEIL	(0 0) 18	11316	729	732	-3	129	R	4
230.	IHEDPHILUS	(0 37) 8	5372	436	440	-4	68		11
231.	SCHWARTZE JUNGE	(0 38) 9	29652	1563	1579	-16	218	R	8
232.	KLEIN UMSTAND	(0 36) 1	0	1	31	-30	1		14
233.	V. HIPPE	(0 41) 6	56000	3657	3688	-31	591		14
234.	FRIEDRICH FETTE HENNE	(37 37) 1	0	0	35	-35	0		13
235.	WULFINGSBURG	(0 0) 18	57532	7590	7642	-52	1088	S I C	13
236.	HAGEN SIEPERBANK	(33 0) 15	72688	6417	6479	-62	882	R I C	6
237.	V. UNTER ST.	(37 38) 2	8000	346	420	-74	43	S	16
238.	SONNENSCHEIN	(35 36) 2	0	0	96	-96	0	R C	6
239.	V. WILDENBERG & VOGEL B.	(0 44) 15	12520	1200	1305	-105	179		11
240.	ALTEMANN S.F.	(42 42) 1	0	0	111	-111	0	R C	6
241.	NICOLAUS	(42 0) 1	0	0	145	-145	0		19
242.	V. SCHIERBANK	(0 0) 18	20320	1883	2029	-146	290		11
243.	FRIEDLICHER NACHBAR	(37 37) 1	0	0	149	-149	0	C	11
244.	FROHE ANSICHT	(0 47) 18	14512	1053	1212	-159	173		2
245.	JOHANN MATHIAS	(36 36) 1	0	0	162	-162	0	C	11
246.	ERBENKAMRSBANK	(0 0) 7	17600	407	571	-164	69		18
247.	TROMPETE	(0 0) 6	31200	1363	1529	-166	262	I	16
248.	ANSELMUS E.	(37 0) 6	30800	1699	1878	-179	317		14
249.	SANDBERG	(42 43) 2	420	21	214	-193	3	C	13
250.	PREOSSISCHER ZEPIER	(39 39) 1	0	0	211	-211	0	C	6
251.	MARGARETHE	(0 43) 14	1452	125	342	-217	26	C	2
252.	IREUE	(0 0) 18	35164	3430	3652	-222	410	R I	6
253.	DIEDERICH	(38 0) 10	36672	4187	4411	-224	688	C	2

	NAME	YEARS OF OPERATION	30-37	OUTPUT	INCOME	COST	LOSSES	TAXES	TECHNOL.	DISTRICT
254.	KRACKHAUSBANK	(39 39)	1	0	0	228	0	680	C	2
255.	HEINRICH	(41 0)	7	11632	877	1130	-253	139	I C	4
256.	NOTTEKAPPE	(0 0)	6	21200	935	1188	-253	178		16
257.	HEINRICH WILHELM	(33 43)	11	15480	1355	1617	-262	210	C	11
258.	VERLORENE POSTEN	(0 39)	10	5256	277	554	-277	51	R	4
259.	VERLOHRNER SOHN	(39 42)	4	5256	277	554	-277	51	C	6
260.	NEUE STEINKUHL	(32 0)	16	30700	2898	3193	-295	458	I	6
261.	CATHARINA LEVERMANNSE	(41 42)	2	9600	1561	1865	-304	185		16
262.	KURZEGGERBANK	(37 39)	3	7056	767	1084	-317	160	R C	10
263.	V. CATHARINA	(0 0)	7	3600	306	647	-341	61	I	16
264.	V. HERMANN	(38 0)	5	105600	6049	6391	-342	989		15
265.	CATHARINE	(34 45)	12	400	109	452	-343	5	R	9
266.	PRINZ FRIEDRICH	(41 0)	2	1200	64	426	-362	0		19
267.	V. HOFFNUNG	(0 0)	7	18400	1102	1472	-370	214		15
268.	WILHELMINE SHCNEPPE	(40 0)	1	5200	206	604	-398	40		19
269.	MAXIMUS	(24 45)	11	22144	2090	2498	-408	320	R	10
270.	GROSSE BOVERMANNSW.	(40 0)	3	16400	717	1129	-412	123		14
271.	JUNGE WITWE	(37 38)	2	0	0	424	-424	0	C	3
272.	V. PAUL V. HIPPE ST.	(0 40)	5	1200	86	519	-433	22		14
273.	SCHULTENKAMPERBANK	(38 39)	2	0	0	440	-440	0	C	1
274.	SCHUTZDELLERBANK	(0 41)	6	22800	1180	1625	-445	195		18
275.	V. CAPPELLENBANK	(0 0)	7	73200	5454	5921	-467	941	I	15
276.	TROMPETE UNTER MARK	(0 41)	6	29600	1300	1772	-472	242	I	16
277.	JULIANE	(39 0)	9	19148	1321	1809	-488	220	I C	4
278.	MUHLE	(0 0)	1	12800	327	815	-488	72		18
279.	FINIGKETT S.F.	(42 0)	6	12984	259	754	-495	41	R C	8
280.	FRISCHES GLUCK & FRIED.	(0 39)	4	6000	240	835	-495	71	I	16
281.	EINTRACHT S.F.	(42 43)	2	12984	259	759	-500	41	R C	8
282.	CAROLINE	(0 0)	18	5228	740	1242	-502	67		2
283.	KLEINGLUCK	(40 42)	3	5320	114	619	-505	25	C	12

	NAME	YEARS OF OPERATION 30-37	OUTPUT	INCOME	COST	LOSSES	TAXES	TECHNOL.	DISTRICT
284.	FERNER GLUCK	(39 40) 2	600	46	558	-512	11	C	11
285.	BETTELSTRUCKER ST.	(0 0) 7	25600	2033	2576	-543	353		17
286.	GLUCKSFORTGANG	(39 40) 2	5912	449	993	-544	74	R C	10
287.	V. DEIMELSBERGER E.	(0 0) 7	11600	932	1497	-565	161	I	17
288.	V. FRIEDRICH WILHELM	(33 44) 12	3695	262	833	-571	54	R C	10
289.	MAASBANK	(37 0) 6	56400	2680	3266	-586	496	I	14
290.	NIPSHAGEN	(0 38) 1	28000	1193	1782	-589	341		14
291.	IGNATIUS	(33 42) 10	16688	1581	2179	-598	177	R I C	6
292.	FEIGENBAUM	(38 44) 7	11512	474	1108	-634	80	C	11
293.	HORTENSIA	(41 44) 4	0	0	650	-650	8	C	6
294.	GUTE KAUF	(37 38) 2	1195	72	749	-677	11	R C	10
295.	GEDOLD	(37 43) 7	6848	582	1362	-680	114	C	11
296.	BEISENKAMP	(0 0) 7	23200	859	1548	-689	163	I	16
297.	PAULINER E.	(0 0) 7	6000	365	1123	-758	64		14
298.	SIEBEN SOHNE	(42 42) 1	2296	165	953	-788	42	C	11
299.	GOLDFINKEL	(39 40) 2	800	17	841	-824	5		17
300.	DER NEUE WEG	(40 0) 8	0	0	832	-832	18	R C	7
301.	STEINGATT	(0 0) 7	155200	14476	15308	-832	2100	S I	16
302.	HEIMANNFELD	(38 0) 5	19200	871	1706	-835	150		16
303.	HALTER	(37 0) 11	0	0	852	-852	0	I C	6
304.	FORTUNA	(40 46) 7	24156	1917	2808	-891	322	R I C	7
305.	ISABELLE	(33 0) 7	1068	53	954	-901	16	C	6
306.	PAUL	(40 0) 3	2400	151	1085	-934	29		14
307.	KALSKSIEPEN	(0 0) 7	35600	2440	3409	-969	138	I	18
308.	FRISCHGEWAGT	(37 0) 11	0	0	979	-979	0	C	13
309.	HARMONIE	(38 0) 10	14264	1111	2095	-984	180	C	11
310.	CHRISTINE H. G. S.F.	(42 44) 3	43512	3772	4756	-984	483	R C	7
311.	ST. MATHIAS	(0 0) 18	55572	6710	7720	-1010	836	R I	7
312.	TROTZ	(0 40) 5	13600	587	1699	-1112	121		18
313.	EINIGKEIT	(34 0) 14	58592	3459	4625	-1166	506	R I C	8
314.	IDUNA	(42 0) 6	0	0	1192	-1192	9	R C	7
315.	V. FIEDRICHSFELD	(30 0) 18	9292	862	2212	-1350	130	R	4
316.	HELENA	(31 0) 17	2164	341	1735	-1394	57	I	4
317.	ELIAS E.	(31 0) 17	22896	2045	3578	-1533	301	I C	11

	NAME	YEARS OF OPERATION 30-37	OUTPUT	INCOME	COST	LOSSES	TAXES	TECHNOL.	DISTRICT
318.	LOUISENGLUCK	(0 0) 18	136204	15343	16894	-1551	2213	S R I C	10
319.	KRUSENBERG	(41 0) 2	80800	2416	4098	-1682	373		14
320.	V. KONIGSCRONE	(38 40) 3	5612	362	2118	-1756	82	C	6
321.	V. LOUISE	(39 0) 3	46800	2503	4356	-1853	441	I	19
322.	GLUCKAUFER E.	(35 0) 13	2020	113	2204	-2091	52	C	3
323.	V. PRASIDENT	(42 0) 6	0	15	2247	-2232	0	S I C	5
324.	AUEGOITEWAGT T.	(36 37) 2	43716	4915	7183	-2268	632	R C	10
325.	PIEPERS E.	(42 0) 6	3352	125	2782	-2657	23	I C	11
326.	TIEFER ROMBERGER E.	(0 42) 12	0	345	3257	-2912	0	C	2
327.	HORSTER E.	(36 0) 12	0	7	3180	-3173	107	R I C	8
328.	DRECKBANKER E.	(41 0) 7	0	28	3396	-3368	41	I C	12
329.	JULIUS PHILIP E.	(39 0) 9	22732	23	3889	-3866	25	R I C	6
330.	V. HOBEISEN	(0 0) 7	88000	8199	12676	-4477	1139	S I	17
331.	AMSEL & SCHNEPPE	(42 0) 1	73600	2238	6758	-4520	356		19
332.	DACHS & GREVELSLOCH & D.	(42 42) 1	0	0	4565	-4564	0	C	12
333.	V. ENGELSBURG	(34 0) 14	220500	20129	25500	-5371	2884	S I C	5
334.	V. CHARLOTTE	(35 0) 13	250603	18044	23883	-5839	1500	S R I C	9
335.	URBANUS	(0 0) 18	113868	9531	16001	-6470	1494	S R I C	6
336.	GLUCKAUE TIEFBAU	(41 0) 7	14524	132	9298	-9166	54	C	3
337.	VEREINIGUNG UBER ST.	(38 0) 5	34400	1080	10984	-9904	196		16
338.	KRONENPRINZ V. PREUSSEN	(0 0) 5	41600	2039	12448	-10409	333	S I	17
339.	MONCHHOFFSBANK	(38 0) 2	2000	97	12602	-12505	30	S	16
340.	V. NACHT & AUFGOTTG.	(40 0) 8	18843	2168	17689	-15521	345	S R I C	10
341.	V. NACHTIGAL T.	(32 0) 16	18843	2167	17688	-15521	345	R I C	10
342.	WOLFSBANK	(38 0) 3	82800	9658	28763	-19105	1028	S	17
343.	V. HAGENBECK	(0 41) 4	15600	294	20177	-19883	55	S	17
344.	FRANZISKA TIEFBAU	(39 0) 9	217404	22589	43779	-21190	2981	S R I C	4
345.	GRAF BEUST	(42 0) 1	253200	32523	85617	-53094	4552	S I	17

Note: The operations missing between 160 and 229 did not show any profit or loss.

Source: Monthly Mining Reports, 1830-47; Annual Mining Reports, 1823-43

TRANSPORTATION MODES & LOCATION OF MINING OPERATIONS

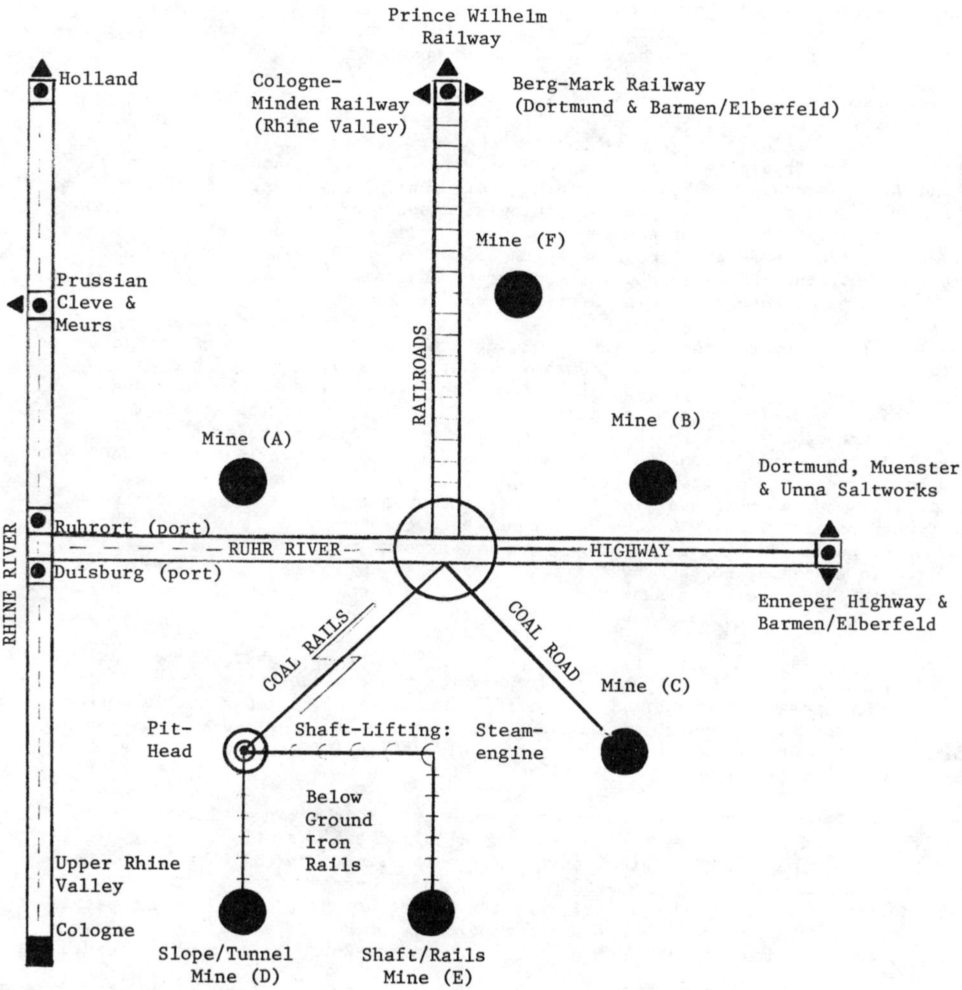

EXPLANATION

The following is a schematic presentation of the location of six typical coal mining operations and their means of transport to major market areas. During the century under discussion the location of mining activity, mining technique, and modes of transportation underwent dramatic changes.

Mining operations A, B, C, were typically located in or near the Ruhr river valley or a major road. They used similiar mining techniques such as open pit, slope or tunnel mining or at best shallow shafts. Operation (A) was located near the Ruhr river transporting coal to the port city of Ruhrort and then to the Rhine Valley and also to Holland. Mining operation (B) was located near one of the major roads or highways. In the mining area south of the Ruhr river coal was transported via the Enneper Road to the Wupper valley market and north of the River via highways to the Unna region and Saltworks as well as to the larger cities of Dortmund and even Münster. Operation (C) gained access to either the River or a major road by buiding a private coal road.

The location of operations D,E,and F were determined by a shift in mining activity, new mining technology and transportation. Earlier coal mining had depleted the more easily accessable coal seams and a new mining technology was required. In addition, mining activity moved further away from the traditional modes of transport. During the second quarter of the nineteenth century, operations such as (D) and (E) increasingly had to use coal rails both above and underground and shafts with steam-engines lifting coal. Operation (D) was typically a tunnel or slope mine built into the side of a hill or one of the many Ruhr river tributary valley. The tunnelbecame longer and longer and carts had to be pushed by miners or horse drawn underground rail systems had to be built. In addition, horse drawn railways were also built above ground from the pithead to the valley of a Ruhr river tributary or the Ruhr itself. These rails were frequently used by a group of mines in the same area.

Operation (E) is similar to (D) except it had to construct a shaft. By the late 1830's these shafts already penetrated the water bearing chalk layer present in the northern part of the Ruhr mining region. The shaft presented a bottle neck in the transportation system without the steam-engine. It required a steam-engine to lift coal. Both operations (D) and (E) could still ultimately market via the Ruhr river or major roads.

Operation (F) similiar mining techniques could transport their coal with the newly developed railroad system. During the 1840's these operations were typically located in the newer mining areas in the far norther part of the Ruhr region near the Emscher valley. During the 1850's some of these operations build their own private railroad links. Of course, with the advent of the railroad, coal operations could market to nearly all market areas previously only accessable by the more cumbersome traditional means of transport such as the Ruhr river or major roads.

TABLE

1836-48, MARK IRON RAILS: Length Above and Below Ground (Lachter*)

Name of Mine	Below Ground 1836	1840	1842	1848	Above Ground 1836	1840	1842	1848
1. V. Hasenw. & Himmelscr.	1100	1275	1285	1619	1250	3046	1696	3126
2. Glueckauf				273				
3. V.M.A. & Steinb.				2599				1044
4. V. General & Erbst.	1800	2180	2180	3130	2600	2300	2292	2547
5. Franzisca Tiefbau			144	4300			350	1500
6. V. Praesident				444				445
7. Friederica	300	1250	1630	3375	160	320	320	320
8. Mecklingsbaenker E.			100	100			117	117
9. Friede		350	435	445				31
10. V. Engelsburg					930	935	999	999
11. V. Nachtigall		1620	3650					
12. V. Nachtigall & Aufgot.			181	6578			138	1015
13. Trappe		400	440	990	SH	SH	SH30	SH60
14. V. Morgenstern Tiefb.		330	320	528		155	155	330
15. Johann Friedrich		1100	1100	1548	H	H45	H79	250
16. V. Charlotte		165	252	937	363	760	760	P1253
17. Eintracht	906	937	1057	834	172	178	189	140
18. Himmelsfuersten E.	900	615	615	725		46	46	46
19. Kirschbaum				250				HH100
20. Urbanus				1247		33		2182
21. Louisenglueck		40	90	562		40	70	267
22. Frischauf				1080				129
23. Louise	-	-	-	500	-	C	C	450
24. Lucia		10	10	10		203	HH203	HH203
25. V. Neue Missgunst	-	53	313	-	G_2	$G_3$252	579	-
26. Morgenstern-Osten	-	(14)	(14)	(14)	-	(14)	(14)	(14)
27. V. Preuss. Adler					GV			GV
28. Carl Friedrich	700	1100	1256	1306	1000	1000	1000	1000
29. V. Stock & Scherenb.		292					1008	552
30. V. Altesackb. & Geitl.		435	425		1100	45	30	
31. Wohlverwahrt		1004	1004				75	75
32. Forelle & Alte Weib						75	75	
33. Storcksbank			W					
34. Gueldene Sonne								
35. Gluecksburg		103	210	230	G_2	G_2G_3	G77	20

1836-48, MARK IRON RAILS: Length Above and Below Ground (Lachter*)

	Below Ground				Above Ground			
Name of Mine	1836	1840	1842	1848	1836	1840	1842	1848
36. Carl Friedrich					700			
37. Sieper & Muehler Gr.			181	30				
38. Muehlheimerglueck	300	325	325	325	23	36	36	36
39. Fortuna-Westen		148	148	148				210
40. Aufgettgewagt		870	2060	2187				
41. Wuelfingsburg			15		SH	SH26	SH26	SH81
42. Stoecker Hauptgr.		324	908	485		10	20	10
43. Dachs & Grevelsloch		102	292	562	SH	SH	SH10	SH257
44. Gleckauf & Segen			140	649				
45. Concordia			336	588			57	74
46. Glueckauf			992	260				15
47. St. Peter					SH	SH		SH110
48. Fortuna			235					
49. Wiederlage		660	1120					
50. V. Anna Cath. & Dickeb.					H	H		
51. Carl Wilhelm	-	-	-	21	-	-	-	330
52. Dannenbaum	-	-	360	-	-	-	-	-
53. Juliane	-	-	-	-	-	-	-	-
54. Hagensieperbank	-	443	924	622	-	$G_3$160	$G_1$160	120
55. Einigkeit	-	330	480	1341	-	124	124	460
56. Wiendahlsnebenbank	-	-	-	128	-	-	-	-
57. Gibraltar Erbst.	250	411	822	657	100	100	100	150
58. Stuecktey	-	280	280	-	-	-	-	-
59. Piepers Erbst.	-	-	84	-	-	-	-	-
60. Altemann	-	130	188	-	-	-	-	-
61. Heinrich	-	-	102	-	-	-	-	290
62. Buntebank	-	-	-	-	-	C	-	-
63. Jupiter	-	-	-	60	-	-	-	-
64. Neue Steinkuhl	-	-	54	-	-	-	150	-
65. St. Matthias	1000	1325	1325	1358	H	H20	H20	640
66. Elias Erbst.	-	-	60	-	-	-	350	-
67. Hammerbank	-	-	498	-	-	-	225	-
68. Treue	300	400	400	-	200	150	150	-
69. Christiane & Hilfe G.	-	-	140	-	1415	HH140	HH	
70. St. Johannes E.	-	(82)	(82)	(82)	-	(82)	(82)	-

Name of Mine	Below Ground 1836	1840	1842	1848	Above Ground 1836	1840	1842	1848
71. Ignatius	-	120	120	-	-	-	120	-
72. Dreckbaenker E.	-	306	1649	-	-	-	16	20
73. Helena	202	362	1160	908	180	150	385	174
74. Clausthal	-	-	-	-	-	450	450	-
75. Julius Philipp	-	30	274	1949	-	14	552	699
76. V. Gr. Varstb. & Altend.	-	997	1017	1017	1000	39	39	39
77. Papenbank	-	-	-	40	-	-	-	H277
78. Halter	-	-	200	-	-	-	15	-
79. Friedrich	-	-	-	564	-	-	-	CH591
80. Carl Wilhelm E.	-	-	-	415	-	-	-	-
81. Hoerster Erbst.	-	-	-	(3)	-	-	-	PW
82. St. Joh. E. & Friehlhs.	-	845	870	900	404	563	563	809
83. V. Nachtigall Tiefb.	-	150	150	-	-	248	646	-
84. Friedrich Wilhelm	-	-	-	45	-	-	-	460
85. Wilhelmine	-	-	-	304	-	-	-	-
86. Prinz Wilhelm Bahn	-	-	-	-	-	-	-	-
87. Schlebusch/Harkerten	-	-	-	-	3600	3708	4377	3768
88. Muttenthaler Baha	-	-	-	-	2500	2500	2500	-
89. Rauendahler Bahn	-	-	-	-	900	925	-	-
90. Gluecksb. & Hagensiep	-	-	-	-	-	-	-	972
91. Gluecksb. & Neue Miss	-	-	-	-	2035	1168	-	-
92. G. & Neue Missgunst & Hagensieperbank	-	-	-	-	-	804	1972	1983

Symbols for Group rails:
SH - (87) PW - (86)
HH - (1) H - (18)
G2 - (90) VM - (3)
G3 - (91) GV - (76)
C - (74)

Source: For 1836, 40, 42, B 38-4
For 1848 No. 603 (N)

TABLE

1836-48, MARK COAL RAILS: Length and Cost

	A. Total Length (ltr.)				B. Total Cost (Rth.)		
	1836	1840	1842	1848	1836	1840	1848
1.	2350	4321	2981	4745	9313	14226	15270
2.	-	-	-	273	-	-	820
3.	-	-	-	3643	-	-	11097
4.	4400	4488	4472	5677	19667	16872	22655
5.	-	-	494	5800	-	-	5292
6.	-	-	-	889	-	-	3249
7.	460	1570	1950	3695	2650	7732	15662
8.	-	-	217	217	-	-	579
9.	-	350	435	476	-	599	792
10.	930	935	999	999	3100	4059	4121
11.	-	1620	3650	-	-	4306	-
12.	-	-	319	7593	-	-	18983
13.	-	400	470	1050	-	1200	2190
14.	-	480	475	861	-	599	792
15.	-	1145	1179	1798	-	2862	5291
16.	363	925	1032	-	3592	2997	12533
17.	1073	1115	1246	974	2605	2770	3200
18.	900	661	661	771	2250	1498	1744
19.	-	-	-	350	-	-	-
20.	-	33	-	3429	-	-	23535
21.	-	80	160	829	-	4640	1309
22.	-	-	-	1209	-	-	3651
23.	-	-	-	950	-	-	1724
24.	-	213	213	213	-	571	572
25.	-	305	892	-	-	710	-
26.	-	-	-	-	-	-	-
27.	-	-	-	-	-	-	-
28.	1700	2100	2256	2306	6693	5600	6149
29.	-	292	1008	552	-	524	922
30.	1100	485	455	-	2053	1053	-
31.	-	-	1079	1079	-	-	2518
32.	-	288	-	-	-	288	-
33.	-	-	-	-	-	-	-
34.	-	-	-	-	-	-	-
35.	-	103	287	250	-	217	569
36.	700	-	-	-	-	-	-
37.	-	-	181	30	-	-	1428
38.	323	361	361	361	641	820	820
39.	-	148	148	148	-	350	876
40.	-	870	2060	2187	-	2083	5468
41.	-	26	41	81	-	39	122
42.	-	334	928	595	-	600	890
43.	-	102	302	829	-	135	1483
44.	-	-	140	649	-	-	1207
45.	-	-	393	562	-	-	1523
46.	-	-	992	275	-	-	623

	A. Total Length (ltr.)				B. Total Cost (Rth.)		
	1836	1840	1842	1848	1836	1840	1848
47.	-	-	-	110	-	-	201
48.	-	-	235	-	-	-	-
49.	-	660	1120	-	-	1610	-
50.	-	-	-	-	-	-	-
51.	-	-	-	351	-	-	1264
52.	-	-	360	-	-	-	-
53.	-	-	-	-	-	-	-
54.	-	603	1084	742	-	1484	1731
55.	-	454	604	1801	-	3200	1303
56.	-	-	-	128	-	-	223
57.	350	511	922	807	574	1259	2017
58.	-	280	280	-	-	746	-
59.	-	-	84	-	-	-	-
60.	-	130	188	-	-	325	-
61.	-	-	102	290	-	-	677
62.	-	-	-	-	-	-	-
63.	-	-	-	60	-	-	173
64.	-	-	204	-	-	-	-
65.	1000	1345	1345	1998	3150	3362	6124
66.	-	-	410	-	-	-	-
67.	-	-	723	-	-	-	-
68.	500	550	550	-	1667	1375	-
69.	1415	140	140	-	5800	375	-
70.	-	-	-	-	-	-	-
71.	-	120	240	-	-	280	-
72.	-	-	412	1669	-	-	3506
73.	400	512	1545	1082	827	1268	4843
74.	-	450	450	-	-	1310	-
75.	-	44	856	2648	-	105	6031
76.	1000	1036	1056	1056	2133	2474	2513
77.	-	-	-	317	-	-	1411
78.	-	-	215	-	-	-	-
79.	-	-	-	1155	-	-	3080
80.	-	-	-	415	-	-	1038
81.	-	-	-	-	-	-	-
82.	404	1408	1433	1709	3124	3287	8545
83.	-	398	796	-	-	1037	-
84.	-	-	-	505	-	-	1126
85.	-	-	-	304	-	-	499
86.	-	-	-	-	-	-	-
87.	3600	3708	4377	3768	36000	12987	13457
88.	2500	2500	2500	-	-	-	-
89.	900	925	-	-	3150	-	-
90.	-	-	-	972	-	-	-
91.	2035	1168	-	-	4952	2793	-
92.	-	804	1972	1983	-	1922	6672
Total	28408	41488	56659	81199	106848	118549	251103

Source: For 1836, 40, 42 B 38-4.
For 1848 No. 603 (N)

TABLE

1848, MARK COAL RAILS: Length, Cost, Power, Destination

Name & District*		(1)	(2)	(3)	(4)	(5)	(6)	(7)	(8)	(9)	(10)
A	Friedrich Wilhelm	497	45	460	505	1,126	39	104	10	M	Highway leading to Dertmund
	Louise Erbst	827	450	500	950	1,724	43	119	10	M	Berg-Mark Railway
B	Glueckauf	427		273	273	820	55	121	10	M	Highway-Dortmund
	Glueckauf & Seges	129		649	649	1,207		55	10	M	Highway-Dortmund
	Wilhelmine & Sterkb.	86		304	304	499		49	4	M	Highway-Dortmund
	Wiendahlsnebenb.	126		128	128	223		52	5	M	Highway-Dortmund
C	Franziska Tiefbau	431	1,500	4,300	5,800	5,292	84	141	8	MH	Ruhr River Depot
	Frischauf	100	129	1,080	1,209	3,651	84	142	6	M	Ruhr River Depot
	Heinrich	23	290		290	676	70		7	M	Shaft Dorothea to Pit Head Depot
	Helena	38	174	908	1,082	4,842		160	8	M	Ruhr River Depot
D	V. Urbanus	350	2,182	1,247	3,429	23,536	72	131	8	MH	Shaft to River Depot
	Gluecksburg (Group)			1,983	1,983	6,692	68	161		M	Group Rail to River Depot
	Gluecksburg	65	20	230	250	570	68		6	M	To Group Rail to River Depot
	Julius Philipp E.	133	699	1,949	2,684	6,032	68	161	10	M	To Group Rail to River Depot
	Hagensieperbank	39	120	622	742	1,731	70		10	M	To Group Rail to River Depot
	Gibraltar Erbst.	180	150	657	807	2,017	80	75	10	M	River Depot
	Carl Friedrich E.	168	1,000	1,306	2,306	6,149	80		10	MH	Tunnel Mouth to River Depot
	Friedrich	93	564	591	1,155	3,080	80		10	MH	Connects to River Depot
	Carl Wilhelm	77	330	21	351	1,264		110	8	M	Tunnel Mouth to River Depot
	Carl Wilhelm E.			415	415	1,037	75		8	M	Connects to Carl Wilhelm
E	V. General Erbst.	437	2,547	3,130	5,677	22,655	76	196	18	MH	Shafts to River Depot
	Hasenw. & Himmelscr.	410	3,126	1,619	4,745	15,220	86	160	8	MH	Shafts to River Depot
	Kirschbaum		100	250	350		80		8	M	Connects to Hasenw. & Himmelscr.
	Lucia	121	203	10	213	571	80		8	M	Connects to Hasenw. & Himmelscr.
	St. Matthias E.	92	640	1,358	1,998	6,124	79	79	8	M	Group Rauhendahler RR River Depot
	Johann Friedrich	20	250	1,548	1,798	5,291	79	155	15	M	Group Rauhendahler RR River Depot
	Papenbank	108	277	40	314	1,410	79	155	15	M	Group Rauhendahler RR River Depot
F	Friederica	393	320	3,375	3,695	5,661	126	179	6	M	Shaft to Depot Witten-Bochum Hwy
	Praesident	301	444	445	889	3,249	75	136	8	M	Shaft to Bochum-Steele Highway
	V. Engelsburg	350	999		999	4,121	126		8	H	Shaft to Bochum-Steele Highway
	Eintracht	250	140	835	974	3,200	73	139	15	M	Tunnel to River Depot
	Wehlverwahrt	121	75	1,004	1,079	2,518	70		15	M	Tunnel to River Depot
	Mecklingsb. Erb.	329	117	100	217	519	70		8	M	Tunnel to River Depot
	Einigkeit	163	460	1,341	1,801	6,303	67	139	10	M	Tunnel to River Depot
	Hoerster Erbst.	376	1,044	2,599	3,643	11,098	84	91	8	H	Group Tunnel to Steele-Vehwinkel RR
	V.M.A. & Seinbank	376	1,044	2,599	3,643	11,098	84	91	8	H	Group Tunnel to Steele-Vehwinkel RR

		(1)	(2)	(3)	(4)	(5)	(6)	(7)	(8)	(9)	(10)
G	Fortuna im Westen	119	148	210	358	876	65		4	M	
	Jupiter	60		60	60	173	87		6	M	
	St. Jeh. & Frieh- lingh.	111	809	900	1,709	8,545	128	173	6	H	Shaft to Herbede Ruhr Bridge
	V. Nachtigall	469	1,015	6,578	7,593	18,982	68	84	16	M	Group to Herbede Ruhr Bridge
	Aufgettgewagt	28		2,187	2,187	5,467	62	80	6	M	Group to Herbede Ruhr Bridge
	V. Morgenstern Tief	212	333	528	861	1,431	55		6	M	Group to Herbede Ruhr Bridge
	Friede	251	31	445	476	791	51		6	M	Group to Herbede Ruhr Bridge
	Louisenglueck	210	267	562	829	1,309	40	45	6	M	Group to Herbede Ruhr Bridge
H	Schlebusch Hark.RR		3,768		3,768	13,456	108		18	H	Enneper Highway
	Trappe	275	60	990	1,050	2,092	73		6	M	Group to Schlebusch Harkerter RR
	St. Peter	54	110		110	201	54		4	M	Group to Schlebusch Harkerter RR
	Wuelfingsburg	64	81		81	121	45		4	M	Group to Schlebusch Harkerter RR
I	Dachs & Grevelsloch	92	15	850	865	1,483		63	4	M	Pithead & Highway to Wuppertal
	Dreckbaenker Erbst.	7	20	1,649	1,669	3,506		63	5	M	Pithead & Highway to Wuppertal
	Stoecker Hauptgr.	30	10	485	495	890		54	3	M	Pithead & Highway to Wuppertal
	Stock & Scherenber- ger	11		552	552	922		54	3	M	Pithead & Highway to Wuppertal
	Sieper & Muehler Gr.	81		30	30	1,428		69	4	M	Pithead & Highway to Wuppertal
	Glueckauf	37	15	260	275	624		69	4	M	Pithead & Highway to Wuppertal
	Concordia	2	74	588	662	1,523		69	4	M	Pithead & Highway to Wuppertal
J	V. Muehlheimer Glueck	34	36	325	361	820		68	8	M	River Depot
	Himmelsfuersten E.	174	46	725	771	1,744		68	8	M	River Depot
	V. Charlotte	475	1,253	937	2,190	12,533	126	192	18	MH	River Depot & Steele-Vohwinkel Hwy
	Altendorfer Erbst.		39	1,017	1,056	2,512		71		H	Underground R used by Group to River

Source: 603 (N)

Mining Districts*

A Dortmund
B Hoerde-West
C Witten-East
D Witten-West
E Dahlhausen
F Stalleicken
G Hardenstein
H Schlebusch
I Sprockheevel South & North
J Altendorf

(1) Output (1,000 Scheff.)
(2) Length of Above Ground (ltr.)
(3) Length of Below Ground (ltr.)
(4) Total Length
(5) Total Cost (Rth.)
(6) Above cost per double ltr.
(7) Below cost per double ltr.
(8) Capacity of wagon
(9) Power used: M- Man; H - Horse
(10) Destination (Direction) of rails

TABLE

1836-40, ESSEN REGION RAILS: Length (Ltr.) and Cost (Rth.)

Name of Mine	Below Ground 1836	1838	1840	Above Ground 1836	1838	1840	Total 1836	1838	1840	Total Cost 1836	1840
Saelzer Neuak	1037	1062	1072	30	30	38	1067	1092	1110	4196	3977
Kunstwerk	418	460	660	0	30	50	418	490	710	1567	1704
Deimelsberg	53	301	320	130	116	116	103	417	436	520	1202
Krenprinz						148			148		500
Schoelerpad		200	250	136	300	275	136	500	525	450	1431
Pfaffenberg	152	152	152	115	115	115	267	267	267	857	873
V. Capellenberg	560	560	560	92	92	92	652	652	652	2243	2133
Baldeneyer E.	755	730	730	0	0	0	755	730	730	2044	1957
Hundsnocken	289	563	741	0	46	165	289	609	906	930	3336
Bruchkaemper St.	0	0	0	45	30	50	45	30	50	158	228
Langenbrahm	557	821	1028	220	331	347	797	1152	1375	3108	5960
Kangesbank		82	176		31	55		133	231		844
Gewalt	388	609	630	269	390	570	657	999	1200	2801	4028
Henriette	402	1196	1580	56	166	166	458	1363	1746	1375	5296
Catharina	198	198	198	48	48	48	246	246	246	776	773
Neuglueck	558	445	500	52	52	52	610	497	552	1775	1600
Peertingssiepen	655	721	984	288	288	296	943	1009	1280	4400	6286
Preussenberbecksie		0	109		84	217		84	326		1159
V. Hebeisen		0	0		1130	1310		1130	1310		3930
V. Hitzberg & Dickeb.	500	556	593	54	54	90	554	610	683	1630	6930
Kalksiepen		3	51		270	246		273	297		797
Steingatt	1347	1239	2087	225	86	126	1572	1325	2213	4580	7259
Beisenkamp	324	290	280	340	336	220	664	626	500	1473	1172
Trempete	25	67	150	5	0	0	30	67	150	63	375
Schwarze Adler	570	514	563	490	500	488	1060	1014	1051	1088	2478
Maasbank		236	193		162	100		398	293		517
Nierenbank			80			122			202		356
Petersberg			100			289			389		2919
V. Louise			183			146			329		1017
V. Sellerbeck	1025	1500	0	360	3360	3210	1385	4860	3210	5224	12992
Wehlgemuth	0			120			120			419	
Frisches Glueck	98	141		78	88		176	229		396	
Wiesche	0	0		0	180		180				
Total	9911	12646	13970	3153	8315	9147	13004	20982	23117	42073	84029
Average	495	527	537	158	320	327	565	749	770	1829	2801

Source: B 38-4

TABLE

1839,MANUFACTURERS OF STEAMENGINES AND THEIR USE

(In Rhenish-Westphalian Mining Region)

	NAME	LOCATION	TYPE OF USE & NUMBER				
	(1)	(2)	(3)	(4)	(5)	(6)	(7)
1.	Kamp & Harkort	Wetter	15	25		25	40
2.	Kamp & Company	Wettet	3				3
3.	Jacobi, HANIEL, Huyssen	Sterkrade	19	13	10	23	42
4.	Dobbs & Relaux	Aachen(Eachweiler)	1		3	3	4
5.	Dobbs & Poengen	Aachen	1				1
6.	Dobbs & Nettschen	Aachen	8				8
7.	Reuchhaus J.	Aachen(Eschweiler)	2				2
8.	Dinnendahl, Nehring, Boegel	Isselburg	3	1	10	11	14
9.	Dinnendahl, Dues Moll	Muehlheim	1		4	4	5
10.	Dinnendahl J.	Muelhlheim	4	3	5	8	12
11.	Dinnendahl F.	Muehlheim(Berghausen)	4		5	5	9
12.	Dues, Mcll,Goering	Muehlheim	5				5
13.	Godwin & Werte	Elberfeld	1				1
14.	Haas & Weidmann	Lennep	2				2
15.	Plagemann	Heorschel	1				1
16.	Cockrill	Luettich			1		1
17.	Foreigh but Unknown		1				1
18.	Fairbain W.	Manchester	1				1
19.	Rendhaus J.	N/N	1				1
20.	Lange	Lennep (Wald)	1				1
21.	Unknown		1				1
		Totals	75	42	38	80	155

(1) Name of Manufacturer
(2) Location
(3) Industrial Operations
(4) In Mark Mining Operations, for lifting & water pumping
(5) In Essen-Werden Mining Operations, for lifitng and pumping
(6) Total number used in Ruhr Mining Operations
(7) Total Number Manufactured by each builder

Source: No. 1186 (N)

TABLE

1814-42, STEAMENGINES IN ESSEN MINING OPERATIONS: Type and Horsepower, Builders

(W - Water Pumping, L-Lifting,W-Both)

NAME	LOCATION	1814	1828	1836	1839	1842	Builders*
Saelzer-Neuak	Essen	L21	W90	W90	W90	L13	4
			W32	W32	W32	W41	12
			L 8	L 8	L 8	W118	4
Gewalt	Hinzel		W44	W44	W44	L21	17
				W44	W144	W157	10
				L19	L19	W157	10
Wohlgemuth	Beyfang	L16		1 8	111		11
Sonnenschein		L34					
Nottenkamperbank							
Hellenbank		L 8					
Kunstwerk	Bergerhausen	L34	W80	W80	W170	L14	12
			W32	1 8	W114	L16	
			L 8	L 8	W52	L29	9
					L 8	W58	12
					1 8	W124	12
Sellerbeck	Eppingshefen		W13	W93	W160	L20	9
			L 9	L15	W93	L29	4
				L 9	L20	W150	5
					L15	W160	9
					L 9		
Wiesche	Heyssen		W40	W40	W132	L16	5
			L10	W40	W48	W16	11
				L10	L22	W60	11
Henriette	Beyfang			W44	W44	L13	4
				L11	L11	W157	4
Hobeisen	Holthausen			L 8	L 8		5
					W 34		11
Portingssienpen	Beyfang			L 8	W*15	W*17	4
Ihlandsbrant	Fischlaken			L 8	W*19	L15	9
Kronprinz v.Preussen	Borbeck			L 9	L 9		4
					W66		4
Wolfsbank	Borbeck				W20	W*20	4
Steingatt	Beyfang				W*11	W*23	10
Langenbrahm	Bredeney			L10	W30	L11	9
Hagenbeck	Fronhausen				L26	W*25	4
Vereingung	Oberruhr				W45	W*45	9
Moenchhofsbank	Oberruhr				W*9	W*24	9
					W*9		9
Hundsnacken	Heisingen				W*9		9
Schoelerpad					W143		11
					L 14		9
Graf Beust						W*25	
Hardenberg						W*24	
Helena & Amalia						L 40	
						W 70	

* For 1839 Builders of Steamengines, see Table "manufacturers"

Source: For 1814&1828,Reuss, Mitteilungen; For 1836&1842,Annual Mining Reports
For 1839, No. 1186 (N)

TABLE

1814,1828,1836, PRICE OF STEAMENGINES: Type, Horsepower & Price
(W - Water Pumping, L-Lifting, W-Both)

NAME	1814		1838		1836	
Saelzer-Neuak	(L121)	12,500	(W90)	23,700	(W90)	23,700
	--	--	(W32)	10,000	(W32)	10,000
	--	--	(L 8)	3,400	(I 8)	4,200
Gewalt	(L)	4,200	(W44)	13,000	(W44)	13,000
	--	--	--	--	(W44)	13,000
	--	--	--	--	(L19)	3,800
Wohlgemuth	(L16)	2,500	--	--	(L 8)	5,000
Sonnenschein	(L34)	6,100	--	--	--	--
Nottenkamperbank	(L)	10,000	--	--	--	--
Hellenbank			(L 8)	1,800		
			(W80)	13,600	(W80)	13,600
			(W32)	11,500		
			(L 8)	5,000	(L 8)	5,000
			(L 8)	5,000	(L 8)	5,000
Sellerbeck			(W13)	3,400	(W93)	20,000
			(L 9)	2,000	(L 9)	3,400
			--	--	(L15)	4,300
			(W40)	15,000	(W40)	15,000
Wiesche			(L10)	4,900	(L10)	4,900
					(W40)	15,000
					(W44)	14,500
Henriette					(L11)	3,000
					(L 8)	5,000
Hobeisen					(L 8)	4,000
Portingssiepen					(L 8)	4,000
Ihlandsbarndt					(L 9)	2,000
Kronprinz v. Preusen						

Sourse: Reuss, Mitteilungen (1814, 1828)
No. 1186(1836)

TABLE

1828 & 1836 COST OF OPERATING STEAM ENGINES

	(1)	(2)	(3)	(4)	(5)
Saelzer-Neuak	W 90	2,590		4,370	23,700
	W 32	3,180		1,660	10,000
	L 8	630		850	3,400
Kunstwerk	W 80	2,570		2,840	13,600
	W 32	3,090			11,500
	L 8				5,000
	L 8				5,000
Gewalt	W 44	2,450		2,610	13,000
			W44	5,240	13,000
			L	1,400	3,800
Sellerbeck	W 13	1,430		1,900	3,400
	L 10	420		640	2,000
			L10	710	3,400
Wiesche	W 40	N/N		2,310	15,000
	L 10	720		1,510	4,600
			W	3,880	15,000
Hellenbank	L 7	630			1,700
Kronprinz v.Preussen	L	1,440			2,000
Henriette			W	3,850	14,000
			L	870	3,500
Hobeisen			L	2,310	5,000
Wohlgemuth			L	1,010	5,000
Poertingsiepen			L	1,080	4,000
Ihlandsbraut			L	760	4,000

For 1828

(L) Type and Horsepower: W-Used for water pumping
L-Used for lifting
(2) Cost of Operation, including wages and maintenance(Rth.)
(Based on 300 Day Year)

For 1836:
(3) Type and Horsepower
(4) Cost of Peration

(5) Initial Price of Steamengine

TABLE

1839, IRON WORKS USING STEAMENGINES: Owners,Location, Horsepower and Builder of Steamengine*

NAME	OWNER	LOCATION	Steam-engine H.p.	Maker
Iron Foundry	Knecht &Lasoin	Bochum(W)	16	(1)
Iron Forge (Hammer	Harkort, F.	Hoerde(K)	15	(4)
Rolling Works	Kamp& Hastergberg	Wetter	45	(1)
PuddlingWorks	Kamp& Hasterberg	Wetter	18	(1)
Blast Furnace-Blower	Kamp& Comp	Wetter	6	(1)
BLAST Furnace-Blower	Kamp&Comp	Wetter	6	(1)
Blast Furnace-Blower	Harkort,F.	Wetter	4	(1)
Blast Furnace-Blower	Primavesi&Comp	Gravendorf(T)	9	(5)
Blast Furnace-Blower	JACOBI&Haniel	Rechklingh.	8	(4)
Iron Foundry	Goering,Dues,Mell	Muehlh	10	(13)
Machine Shop	Goering,Dues,Moll	Muehlh.	18	(13)
Iron Works-Steam Engine	Nehring,Boegel,Dinn.	Isselburg	6	(9)
Iron Foundry	Stauber,J.	Elberfeld	8	(11)
Iron Foundry	Godwin& Werte	Elberfeld	6	(14)
Iron Foundry	Haas,F.	Lennep	10	(15)
Blast Frunace-Blower	Haniel&Huyssen	Sterkrade	30	(4)
Blast Frunace-Blower	Haniel&Huyssen	Sterkrade	12	(4)
Machine Ship	Haniel&Huyssen	Sterkrade	30	(4)
Puddling Works	Jacobi,Haniel etc.	Oberhausen	45	(4)
Rolling Works	Jacobi,Haniel etc.	Oberhausen		(4)
Cast-Iron(Steel) Works	Krupp,WW.F.	Essen	27	(4)
Steam Blower &Grinding	Nehring,Boegel,Dinn.	Essen	9	(9)
	Total Horsepower		338	
	Average		15	

*For Name of Builder, see Table on "Manufacturer of Steamengines"

Source: No. 1186(N)

TABLE

1839, TEXTILE OPERATIONS: Owners, Location,& Use of Steam-engines

NAME	OWNER	LOCATION	Steam-engine H.p.	Maker
Spinning Mill	Ossak & Buerther	Bochum(W)	16	(1)
Spinning Mill	Oberempt,J.A.	Hagen(R)	16	(1)
Linen Cloth Finishing	City (Pub)Owned	Bielefeld	6	(4)
Cloth Finishing-Bleaching		Muenster(W)	10	(1)
Cloth Factory	Schmachtenberg	Muelheim	5	(13)
Cloth Factory	Hartman	Dusseldorf	12	(1)
Dying Printing Fabric	Deus&Moll	Dusseldorf(P)	10	(10)
Cotton Spinning Mill	Peill	Elberfeld	10	(1)
Linen Finishing	Gebr.Frowein	Elberfeld	3	(1)
Silk and Velvet	Bockmuehle & Bro.	Elberfeld	12	(18)
Weaving Mill	Frowein,F.	Elberfeld	3	(11)
Silk Factory	Kehler,H.	Elberfeld	12	(3)
Yarn Factory	Beddinghaus	Elberfeld	8	(3)
Silk Factory	Ritterhaus	Barmen	4	(6)
Silk Dying Factory	Heimendahl	Barmen	6	(3)
Cotton Spinning Mill	Wittenstein	Barmen	26	(7)
Cotton Spinning Mill	Jung,F.	Barmen	36	(19)
Cloth Factory	Reichmann	Lennep	7	(1)
Cotton Spinning Mill	David,P.	Lennep	12	(7)
Woollen Cloth Factory	Engels&Oelbermann	Lennep	12	(8)
Woolen Cloth Facotry	Wather,F.	Lennep	10	(7)
Woollen Cloth Facotry	Weulfing,J.	Lennep	36	(&)
Woollen Cloth Facotyr	Schnabel Gebrueder	Lennep	20	(4)
Woollen Cloth Factory	Bockhacker,C.	Lennep	12	(4)
Woollen Cloth Factory	Bauendahl,A.	Lennep	10	(7)
Woollen Cloth Factory	Bauehdahl,A.	Werden	16	(7)
Spinning And W-aving	Thuring	Werden	12	(7)
Woollen Cloth Factory	Strohm	Werden	8	(4)
Woollen Cloth Factory	Schuermann	Werden	20	(12)
Woollen Cloth Factory	Haas,W	Werden	12	(12)
Woollen Cloth Factory	Teschemacher	Werden	14	(12)
Woollen Cloth Factory	Fortmann&Hoffmann	Werden	4	(12))

Total Horsepower 456
Average 13

Steam-engines-Maker, see Table on Manufacturing of Steamengines
Source: No. 1186(N)

TABLE

1790's to 1850's OWNERS OF MINING AND OTHER INDUSTRIAL ENTERPRISES AND THEIR CREDITORS

A. Owners - Their Debt, Occupation & Residence

	DEBT	YEAR	# OF MINES	OWNER	OCCUPATION	RESIDENCE
1.	2,450	1831	1	Albritz		
2.	200	47	1	Assmann E.	Stonemason	Barmen
	180	47				
	150	52				
	4,000	54				
	3,200	55				
3.	1,900	17	4	Bankharts P.		
4.	11,000	34	2	V. D. Becke F.		Hemer
	27,716	40	4			
5.	2.150	16	3			
6.	4,000	42	1	Beisenhertz F.		Duelmen
	4,000	43				
	1,800	43				
	4,533	43				
	429	45				
	1,500	46				
7.	2,700	16	42	Berger Fam.	Widow	
	8,000	26				
	18,000	48				
	20,000	29	16	Berger C.		Witten
	3,000	49				
	5,000	53				
	3,000	16	22	Berger P.	Mining Official	
8.	6.000	43	25	Bilstein P.		
9.	1,850	11	4	Bockmueller J.		
10.	10,333	45	2	V. Born J.		
11.	16,000	45	1	Brand A.	Widow	Witten
	25,000	57				
	2,000	45				
	2,000	45	1	Brand G.		Witten
	3,000	49	1	Brand R		Rahlenbecke
	1,000	50				
	4,000	48	1	Brand W.		Ruhrort
12.	6,000	42	1	Brockel C.	Med. Doctor	Gelsenkirchen
13.	6,000	45	2	Brockelmann F.		Neheim
	6,000	57				
14.	2,500	47	1	Broelemann T.	Widow	Dortmund
15.	4,200	26	1	Bruselmann D.		
16.	2,000	17	3	Buchholz C. & E.		
	600	17				
	450	23				
17.	865	34	1	Bussmann	Colonel	Dueren
	4,200	35				

	DEBT	YEAR	# OF MINES	OWNER	OCCUPATION	RESIDENCE
18.	2,360	44	1	Carle H.		Elberfeld
19.	1,100	27	1	Cremer C.		
	1,000	31				
	3,769	33				
20.	3,200	42	1	Dickerhoff		Voerde
21.	3,000	26	1	Droste	Med. Doctor	
22.	3,500	44	10	V. Eickes F. & Lohr J.	Estate Owner	Sprockhoevel
23.	300	13	3	V. Elverfeld L.	Head of County	
	145	24				
	3,000	18	1	V. E. Leo	Farmer	
	3,300	18				
	3,894	24	7	V. E. L.	Lieutenant & Estate Owner	Steinhausen
	5,000	39				
	2,000	47				
	4,000	37				
	1,000	33				
	2,500	33				
	2,000	35				
	40,000	47				
	20,000	47	4	V. E. Ludwig		
	60,000	49				
	22,000	50				
	2,000	50				
	3,760	50				
24.	2,000	47	1	Endemann W. & Vennemann H.	Widow	Bochum
25.	536	21	3	Fermeer J.	Accountant	
	2,250	28	2	Fermeer H.		Steele
26.	500	28	2	Flehinghaus		
	4,000	41				
	100	48				
27.	3,000	47	1	Frantzen		Hoerde
28.	2,500	42	5	Friederich Fam.	Gvt. Accountant	Wiesbaden
	10,000	43				
	2,950	46				
29.	8,000	29	30	Frielinghaus F.		
	1,000	34				
	16,000	36				
	2,800	41				
	1,000	35	12	Frielinghaus J.		
	5,000	40				
	6,000	41	7	Frielinghaus J. H.		Dortmund
	5,000	40				
	8,000	47				
	2,800	41	21	F. Oberste	Widow	
	45,000	49				
	3,000	49				
	1,291	49				
30.	3,000	28	9	Geisler C.	Tax Collector	
	500	31				
	1,000	32				
	2,000	34				
31.	2,010	35	3	Gieseler C.		Hattingen

	DEBT	YEAR	# OF MINES	OWNER	OCCUPATION	RESIDENCE
32.	3,976	38	6	Grafe H.		Zippe
	160	38				
33.	10,423	41		Gute Hoffnung	Iron Work	
34.	26,667	48	8	Harkort F.		Witten
35.	5,500	09	3	Hasenack		
36.	2,000	45	4	Hegemann C.		Aude
	220	46				
	5,000	46	1	Hegemann H.	Court Official	Hamm
	6,000	47				
37.	4,500	26	50	Heintzmann A.		
	3,000	26	3	Heintzmann F.	M.O. Accountant	
	23,076	24	50	Heintzmann J.	High M. Official	
38.	3,000	45	4	Hemmsoth	Dyer	Witten
39.	2,500	49	1	V. Hoevel C.	Widow	Dortmund
	2,750	42	2	V. H. W.		Dortmund
	2,250	42				
	2,500	43				
	5,500	15				
	3,500	47				
	605	47				
	2,000	48				
	5,500	49				
40.	5,000	42	1	V. Hymmen R.	Nobility	Duesseldorf
41.	2,000	46	1	Jung G.		Wicklinghausen
42.	200	50	4	Kaeseler M. &		
	4,150	50		Veerhoff A. &		
	20,000	52		Heidsieck C. &	Lawyer	
	5,000	53		Mueller G. &	Merchant	
	19,400	53		Marmelstein	Innkeeper	
	3,000	53				
	1,500	54				
43.	13	41	22	Kagenbusch P.	Clothprinter	Wetter
	36	41				
	3,000	42				
	1,100	44				
	24,000	49				
	8,000	50				
	1,000	50				
	270	50				
	5,000	41				
44.	1,000	48	9	Klewitz J.		Ruhrort
	5,000	49				
	500	50				
	1,523	50				
	1,575	50				
	1,250	50				
	600	50				
	1,000	48				
	1,515	50				
	1,700	54				
	5,160	53				

	DEBT	YEAR	# OF MINES	OWNER	OCCUPATION	RESIDENCE
45.	2,992	45	3	Koenig F.	Pastor	Witten
46.	2,000	36		Kruse C.		
47.	2,000	47	22	Kueper		Dortmund
	4,625	34	27	Kueper F.		Dortmund
	7,543	34				
	15,000	34				
	1,200	39				
	4,625	38				
	8,050	23	31	Kueper J.		
	2,000	28				
	4,543	30				
	9,000	34				
	4,000	36	4	Kueper L.		
48.	2,000	47	27	Langewiesche A.		Mollenkotten
	3,052	47				
	3,392	47				
	2,000	48				
	1,500	42	20	Langewiesche H.	Educated Farmer	Mollenkotten
	3,216	46				
	3,000	41	15	Langewiesche J.	Brewer	Schwelm
	4,000	43				
	6,266	42				
	2,700	53				
	3,039	53				
	1,800	57				
49.	9,350	17	17	Lahmann C.		Ruhrort
50.	5,000	44	4	Liebrecht C.		Ruhrort
	2,400	35	25	Liebrecht F.		Sundern
	2,400	40				
	1,500	46				
	3,076	28	26	Liebrecht M.	Merchant	
	1,800	31				
51.	2,035	35	5	Lohmann C.	Shoemaker	Ruhrort
	600	41	16	Lohmann F.	Widow	Witten
	2,700	21				
	1.000	43				
	800	49				
	600	57				
	2,000	40	2	Lohmann F.		
	8,400	39	11	Lohmann J.		Ruhrort
	2,845	41				
52.	3,440	47	1	Mallinecki	Widow	Dortmund
53.	3,000	23	1	Meyer		
54.	319	41	10	Mollmann J.	Widow	
	3,923	45				
	650	52				
55.	3,000	41	2	Morgenstern O.	Mining Company	
56.	4,000	36	Muehlen			
57.	4,000	07	10	Mueser		
	4,336	07				
58.	3,000	42	6	Nied. Frielingh		Bommern
	1,400	44				
	8,000	47				

	DEBT	YEAR	# OF MINES	OWNER	OCCUPATION	RESIDENCE
59.	28,216	44	1	Neimann F.		Horst
60.	4,600	30	1	Oberkinkhaus		
61.	2,000	40	1	Pauls Fam. J.		
	6,100	51				
62.	10.000	46	1	Post F.		Eilpe
	3,200	38	14	Post J.		Eilpe
	1.000	38				
	4,000	44				
	10.000	46				
	2,000	49				
	3,016	50				
63.	30,000	48		V. Praesident	Mining Company	
64.	3.000	43	1	Preissler C.		Werden
	500	46				
	359	51				
	500	53				
	550	54				
65.	20.500	48		Prinz Rudolph	Iron Works	Duelmen
66.	5,000	44	40	Quiper G. & R.		Dortmund
67.	1.600	28	35	V. Rappardt Fam.	Nobility	
	13,500	37				
	10,000	30	35	V. R. D.	County Assessor	Potsdam
	4,050	42	2			
			2	V. R. C. T.	Head Forester	
	3,200	38	35	V. R. F.		
	1,100	40				
	4.050	42				
	3,000	42				
68.	3,500	47	1	Riema H.	Merchant	Duisburg
69.	2.000	44	10	Rocholl C.		Luenen
70.		41	7	V. Romberg Frh.	Chamberlain	Bruenninghausen
71.	6,000	42	7	Ruhrmann A.		Bommern
72.		35	10	Rupe J.		
	50.000	37				
73.	3,756	39	11	Scharwaechter L.		
	538	46				
74.	47	35	5	Schee	Widow	
	2,700	33	8	Schee J.		Elberfeld
75.	6.000	46	3	Schorndorff J.	Ruhr Tolls Account	Muehlheim
76.	2,110	46	4	Schuck	Farmer	Beule
77.	4,000	47	1	Schulte F.		Dortmund
	2,550	47	7	Schulte W.		Wengern
	126	49				
	1,360	50				
	413	50				
78.	13,000	57	1	Schultz J.	Merchant	Cologne
79.	7,750	37	2	Schultz F.		
80.	20,000	45		W. Schuerb & Ch.	Mining Company	
81.	1,800	47	1	Schroth H.	City Collector	Sprockhoevel

	DEBT	YEAR	# OF MINES	OWNER	OCCUPATION	RESIDENCE
82.	500	99	17	Speunemann Fam.		Sprockhoevel
	840	24	4	Spennemann S.	Shoemaker	
	500	99	15	Spennemann H. R.		Zur Obergethe
	2,000	21				
	500	99	8	Spennemann J. R.		
	500	99	12	Spennemann H.		
	2.000	21				
	500	99	12	Spennemann J. P.		Sprockhoevel
	520	28				
	2,000	21				
	500	99	5	Spennemann C. S.		
	500	99	5	Spennemann M. F.		
	500	99	5	Spennemann M. S.		
	500	99	9	Spennemann M. C.		
	2,000	21				
83.	10,260	57	1	Springorum	City Counselor	Herdicke
84.	6,923	30	1	Stinnes K.	Merchant	
85.	1,800	33	2	Stoering J.		A. D. Rosendahl
86.	5,900	47	2	Stock D.	Merchant	Sprockhoevel
87.	6,000	46		Stoecker Hpt.	Mining Company	
88.	1,320	02	1	Stueckmann G.	Colonel	Berghofen
	6,233	27				
	4.100	27				
	620	53				
89.	2,000	23	1	Thurn		
90.	2,000	45	1	Trost F.		Herdecke
91.	15,000	31	12	V. Vincke	President of Westf.	Muenster
92.	1,906	34	1	Vosskuhl C.	Merchant	Hattingen
93.	1,000	36	5	Voss F.	Estate Manager	
	16,500	45	6	Voss P.	Med. Doctor	
	1,000	36	15			
	20.000	45				
	1,000	36	7	Voss W.	Estate Owner	Achternberg
	16,500	45				
94.	3,800	44	4	Wegermann P.		Lennep
95.	3,700	57	4	Weinberg S.	Merchant	Wattenscheidt
	10,000	57				
96.	12,916	37	2	Wehrenbold C.		
97.	2,400	34	3	Weischede		
98.	2,250	24	1	Westphal F.		
	2,000	41	1	Westphal J. F.		
	7.000	39	8	Westphal J. J.		
	8,672	41				
99.	500	28	4	Weyerhaus F.		Duesseldorf
	1,000	01	4	Weyerhaus F. W.	Med. Doctor	Duesseldorf
	3,000	21	4	Weyerhaus J.	Widow	Duesseldorf
	1.500	24				
100.	3,800	44	3	Winkelmann T.		Lennep
101.	1,100	20	8	Wiesmann J.		Hattingen
	23,076	25				
	2,000	43	1	Wiesmann J. G.	Widow	Hattingen

	DEBT	YEAR	# OF MINES	OWNER	OCCUPATION	RESIDENCE
102.	2,000	49	1	Wormstall C.		Hoerde
103.	17,947	37	1	Zimmermann	Gvt. Official	

B. Creditors - Their Occupation and Residence

	Creditor	Occupation	Residence
1.	Bros. J. & M. Herz Sohn	Merchants	Schelm
2.	Bartels	Merchant	Barmen
	Voss J.	Baker	Hattingen
	Bartels	Merchant	Barmen
	Sondermann	Farmer	Oberelfringhauser
	"	"	"
3.	N/N	N/N	N/N
4.	Jaeger J.	Pensioner	Iserlohn
	Relatives	N/N	N/N
5.	Wickelhausen P.	N/N	N/N
6.	Lindenk & Olfers	Merchants	Muenster
	Huck H.	N/N	Herdicke
	Timmermann J.	N/N	Wesel
	Prinz Rudolph	Iron Works	Duelmen
	"	"	"
	Schuerning B.	Merchant	Duelmen
7.	N/N	N/N	N/N
	Mark E.	Banker	Hamm
	Eleanore of Preussen	Princess	N/N
	Baam	N/N	Bommern
	Braschmann	N/N	Hamm
	Helmich F.	Mining Engineer	N/N
	N/N	N/N	N/N
8.	Spannagel	N/N	Hoerde
9.	Schulz	Innkeeper	Ritterhausen
10.	Gockel F.	N/N	Luener Brunnen
11.	Braselmann E.	Banker	Schwelm
	Dotthoff	N/N	Schwelm
	Braselmann E.	Banker	Schwelm
	"	"	"
	Marks E.	Banker	Hamm
	"	"	"
	Kingholz I.	N/N	Ruhrort
	Berger C.	N/N	Witten
12.	Schraegmueller A.	Lieutenant	N/N
13.	v. Esseln	Head of County	Soest
	Wuppermann	Estate Manager	Dortmund
14.	Wortmann G.	N/N	N/N
15.	Winkelhausen P.	N/N	Barmen
16.	Luettringhausen	N/N	Elberfeld
	Moellmann	Merchant	Vollmarstein
	Schneider	Pastor	Schwelm
17.	Spennemann C.	N/N	Witten
	Horlbeck P.	Farmer	Bedenscheid
18.	Firmwich	N/N	Cologne
19.	v. Esseln	Head of County	Soest
	Waldhausen W.	N/N	Essen
	Cremer	Mining Official	N/N
20.	Schulz F.	N/N	Luenen

21.	Sethe	High Judge	Berlin
22.	Loehr F. & A.	Distillers	N/N
23.	Haarmann	N/N	N/N
	Lehmann D.	N/N	N/N
	N/N	Gvt. Official	Berlin
	Berger P.	N/N	N/N
	Giesenberg	Widow	N/N
	Borgemeister H.	Widow	N/N
	Marks E.	Banker	Hamm
	Henke	Head of County	Hartrop
	Hoevel	N/N	Z. Ruhr
	Heintzmann	Gvt. Attorney	Hamm
	Cappell	Gvt. Attorney	N/N
	Kersten H. & J.	Merchants	Elberfeld
	v. d. Heydt & Kersten	Merchants	Wuppertal
	Schaffhausener	Bank	Cologne
	v. Heilmann B.	N/N	Krefeld
	Waldhausen M.	Merchant	N/N
	Mining Treasury	N/A	Bochum
24.	v. Crone M.	N/N	Matena
25.	Voss F.	Earlier Owner	Steele
	Gele	Landowner	N/N
26.	N/N	N/N	N/N
	Relatives	N/N	N/N
	Becker J.	N/N	Sprockhoevel
27.	Waltrabe	Merchant	Hoerde
28.	Schulz F.	N/N	Luenen
	Fischer J.	Banker	Barmen
	Roehr F.	Merchant	Wiesbaden
29.	Mark E.	Banker	Hamm
	Sethe	High Judge	Berlin
	Mining Treasury	N/A	N/N
	Liebrecht	Gvt. Attorney	Bochum
	Relatives	N/N	N/N
	Mining Treasury	N/N	N/N
	Oppenheimer S. & G.	Banking House	Cologne
	Mining Treasury	N/A	N/N
	Mark E.	Banker	Hamm
	Liebrecht	Gvt. Attorney	Bochum
	Relatives	N/N	N/N
	Puttenberg + Fam.	N/N	N/N
	Moeller	Merchant	Schwelm
30.	Keneken	Merchant	Barmen
	Koefler F.	Preacher	Langenberg
	N/N	M.O. Secretary	N/N
31.	Tabaullot	Jr. Attorney	Hamm
32.	N/N	N/N	N/N
	J. & M. Herz Sohn	Merchants	Schwelm
33.	Brenner P.	N/N	Altenhagen
34.	de Beaufort	Noble	Gorz, Belgium
35.	v. Eicken	Noble	N/N
36.	Kantz	N/N	N/N

	Hammacher W.	N/N	Dortmund
	Laar	Widow of Pastor	Guetersloh
	Mining Treasury	N/A	N/N
37.	N/N	N/N	N/N
	Regenhertz	Judge	Hamm
	v. Elverfeld Fam.	N/N	N/N
38.	Sethe	High Judge	Berlin
39.	Schulz F.	Merchant	Luenen
	v. Olfers J.	Banker	Muenster
	Thuning J.	Banker	Muenster
	Ostrup B.	Innkeeper	Duelmen
	Schmedding & Sons	Bankers	Muenster
	Mark E.	Banker	Hamm
	Dinnendahl J.	N/N	N/N
	Prinz Rudolph	Iron Works	Duelmen
	Havighorst	N/N	Muenster
40.	Schraegmueller A.	Lieutenant	N/N
41.	Kotthaus	N/N	Cologne
42.	Kappell F.	Widow of Lawyer	Arnsberg
	Deichmann	Gvt. Accountant	Cologne
	v. Olfers H.	Banker	Muenster
	Kaufmann J.	Merchant	Hamm
	"	"	"
	Kaeseler L.	N/N	N/N
	Overwey & Bunger	Merchants	Dortmund
43.	Mining Treasury	N/A	N/N
	Court Retirement	Fund	Hagen
	Niermann H.	N/N	Wetter
	Moellmann & Kissing	Merchants	Iserlohn
	"	"	"
	Bros. Fischer	Bankers	Barmen
	v. Heydt & Co.	Cast Iron Work	Elberfeld
	Stark L.	Manufacturer	Iserlohn
	Tabe H.	N/N	Seppenrade
44.	Waldthausen A.	Merchant	Essen
	Daber C.	Merchant	Muehlheim
	Nottmann W.	N/N	N/N
	Wenk J.	N/N	Buehl
	Dauphin A.	Merchant	Rheinbischofsheim
	Mester B.	Merchant	Warendorf
	Oeppe W.	Merchant	Lippstadt
	Haamen & Steinberger	Merchandise Firm	Cologne
	Dauphin A.	Merchant	Rheinbischofsheim
	Gretzinger C.	Merchant	Mainz
	Cemper T.	Merchant	Cologne
45.	Relatives	N/N	N/N
46.	N/N	N/N	N/N
47.	Kalthoff	N/N	Neuwied
	Brinkmann	Gvt. Attorney	Schwelm
	Kalthoff	N/N	Neuwied
	M. Herz Sohn	Banker	Schwelm
	Brinkmann	Gvt. Attorney	Schwelm
	"	"	"

	N/N	N/N	N/N
	Wischelhausen P.	N/N	Frankfurt A.M.
	N/N	N/N	N/N
	N/N	N/N	N/N
	Liebrecht	Gvt. Attorney	Bochum
48.	Relatives	N/N	Barmen
	Liebrecht	Gvt. Attorney	Bochum
	Braselmann	Banker	Schwelm
	"	"	"
	Liebrecht	Gvt. Attorney	Bochum
	Langewiesche C.	Merchant	Schwelm
	Liebrecht	Gvt. Attorney	Bochum
	Braselmann E.	Banker	Schwelm
	Relatives	N/N	N/N
	Braselmann E.	Banker	Schwelm
	Liebrecht A.	N/N	Sundern
	Langewiesche A.	N/N	Mollenkotten
49.	Waldhausen G.	Merchant	Essen
50.	Liebrecht	Gvt. Attorney	Bochum
	Hasenclever	Gvt. Attorney	Martfeld
	N/N	N/N	N/N
	N/N	N/N	N/N
	Liebrecht L.	Gvt. Attorney	Arnsberg
	Liebrecht E.	Merchant	Ruhrort
51.	Lohmann J.	Merchant	Ruhrort
	Liebrecht H.	Gvt. Attorney	Bochum
	Buergers	Widow	N/N
	Liebrecht	Gvt. Attorney	Bochum
	Relatives	N/N	N/N
	Grimberg	Judge	Bochum
	Liebrecht	Gvt. Attorney	Bochum
	Relatives	N/N	N/N
52.	Mark E.	Banker	Hamm
53.	N/N	N/N	N/N
54.	Relatives	N/N	N/N
	Bros. Fischer	Bankers	Barmen
	Cramer	Mining Official	Vollmarstein
55.	Berghaus T.	N/N	N/N
56.	Muthmann D.	N/N	N/N
57.	Liebrecht	Retired Judge	Sundern
	Kayserfeld	Colonel	Bischofsheim
58.	Liebrecht	Gvt. Attorney	Bochum
	Hiby W.	Tanner	Sprockhoevel
	Mark E.	Banker	Hamm
59.	Relatives	N/N	N/N
60.	N/N	N/N	N/N
61.	Spannagel C.	N/N	Hoerde
	Schmackis	N/N	Wecklinghausen
62.	Anders H.	Banker	Elberfeld
	Jansen	N/N	N/N
	v. Rappardt	Judge	N/N
	Mark E.	Banker	Hamm
	Anders A.	Banker	Elberfeld

	Mark E.	Banker	Hamm
	Mining Treasury	N/A	N/N
63.	Welfare Fund	N/A	Muenster
64.	Horstmann J.	N/N	N/N
	Waltrabe	Merchant	Hoerde
	"	"	"
	Schulz F.	Merchant	Luenen
	Waltrabe	Merchant	Hoerde
	"	"	"
65.	v. Croy A.	Duke	Duelmen
66.	Schmedding	Banker	Muenster
67.	N/N	N/N	N/N
	v. Rappardt	Judge	Paderborn
	v. Fuerstenberg	Baron	Stammheim
	Schulz F.	N/N	Luenen
	N/N	N/N	N/N
67.	Kahlenbeck C.	Estate Manager	Luenen
	Voerst L.	N/N	Unna
	Schulz L.	N/N	Luenen
	Sohn & Frank	Merchants	Luenen
68.	Tietz A.	Pensioner	Duisburg
69.	Thusing & Olfers	Bankers	Muenster
70.	N/N	N/N	N/N
71.	Bros. Fischer	Bankers	Barmen
72.	Engels	Merchants	Barmen
	Schaffhausener	Bank	Cologne
73.	N/N	N/N	N/N
	Relatives	N/N	N/N
74.	N/N	N/N	N/N
	N/N	N/N	N/N
	Herz Sohn J.	Merchant	Schwelm
75.	v. Stueben F.	Official	Arnsberg
76.	Flashoff C.	N/N	N/N
77.	Wehrenbold	Iron Work Director	Luenen
	Huelsenbeck G.	N/N	Schwelm
	Wirminghaus P.	N/N	Sprockhoevel
	Baltz M.	Merchant	Bochum
	Lehnke	Mining Foreman	Dahlhausen
78.	v. Windelstadt	N/N	Cologne
79.	Gockel	N/N	Luenerbrunnen
80.	Mining Treasury	N/A	N/N
81.	Iving F.	N/N	Leveringhausen
82.	Mining Treasury	N/A	N/N
	N/N	N/N	Sprockhoevel
	Mining Treasury	N/A	N/N
	Klein J.	Merchant	Schwelm
	Mining Treasury	N/A	N/N
	"	"	"
	Wiesmann J.	Merchant	Hattingen
	Mining Treasury	N/A	N/N
	Klein J.	Merchant	Schwelm
	Mining Treasury	N/A	N/N
	Muehlen H.	N/N	N/N

	Klein J.	Merchant	Schwelm
	Mining Treasury	N/A	N/N
	"	"	"
	"	"	"
	"	"	"
	Klein J.	Merchant	Schwelm
83.	Hertmann O.	Widow	Oespel
84.	Roeder H.	Judge	N/N
85.	de Werth P.	Pensioner	N/N
86.	Relatives	N/N	N/N
87.	v.d. Krone	N/N	N/N
88.	N/N	N/N	N/N
	Heernmucker	Merchant	Wetter
	"	"	"
	Wecker	Merchant	Dortmund
89.	Liebrecht	Gvt. Attorney	Bochum
90.	Braselmann E.	Banker	Schwelm
91.	Vogt & Hardt H.	N/N	N/N
92.	Wiesmann J.	Merchant	Hattingen
93.	Voss F.	Widow	Essen
	Voss F. E.	N/N	N/N
	Voss F.	Widow	Essen
	v. Vincke	President of Westph.	Muenster
	Voss F.	Widow	Essen
	Voss E.	Estate Manager	Essen
94.	Kaesch F.	Merchant	Lennep
95.	Moenting H.	Merchant	Gelsenkirchen
	Spiegel M.	N/N	Witten
96.	Gockel F.	N/N	Luenerbrunnen
97.	Huecking F.	N/N	Altena
98.	N/N	N/N	N/N
	Walthausen G.	N/N	Essen
	Ruhr Treasury	N/A	Muehlheim
	Oppenheimer S.	Bankinghouse	Dortmund
99.	Agats C.	High Mining Official	N/N
	Kane & Mark B.	N/N	N/N
	N/N	N/N	N/N
	Hasenclever	Gvt. Attorney	Martfeld
100.	Kaesch F.	Merchant	Lennep
101.	N/N	N/N	N/N
	N/N	N/N	N/N
	Relatives	N/N	N/N
102.	Wulff W.	Transporter	Hoerde
103.	Mark E. & Zahn M.	Merchants	Hamm

For explanation of Source, see introduction to appendix.

Source: Debt Books

1790's to 1859, Amount and Frequency of Loans in the Mark*

Years	No. of Loans	Total Debt	Below 5,000	Between 5 - 10,000	Above 10,000
Pre- 1814	39	36,000	30,000	6,000	-
1815 - 19	21	36,000	27,000	9,000	-
1820 - 24	36	50,000	42,000	8,000	-
1825 - 29	51	83,000	46,000	14,000	23,000
1830 - 34	83	150,000	65,000	32,000	53,000
1835 - 39	105	242,000	83,000	35,000	123,000
1840 - 44	148	317,000	162,000	79,000	76,000
1845 - 49	141	507,000	139,000	74,000	294,000
Totals	624	1,421,000	624,000	257,000	569,000

Source: Data is based on the Mark Debt Book, yet many transactions are included from the Essen region.

1790's-1850 RUHR (MARK) MINING DEBT: AMOUNT AND NUMBER CF TRANSACTIONS

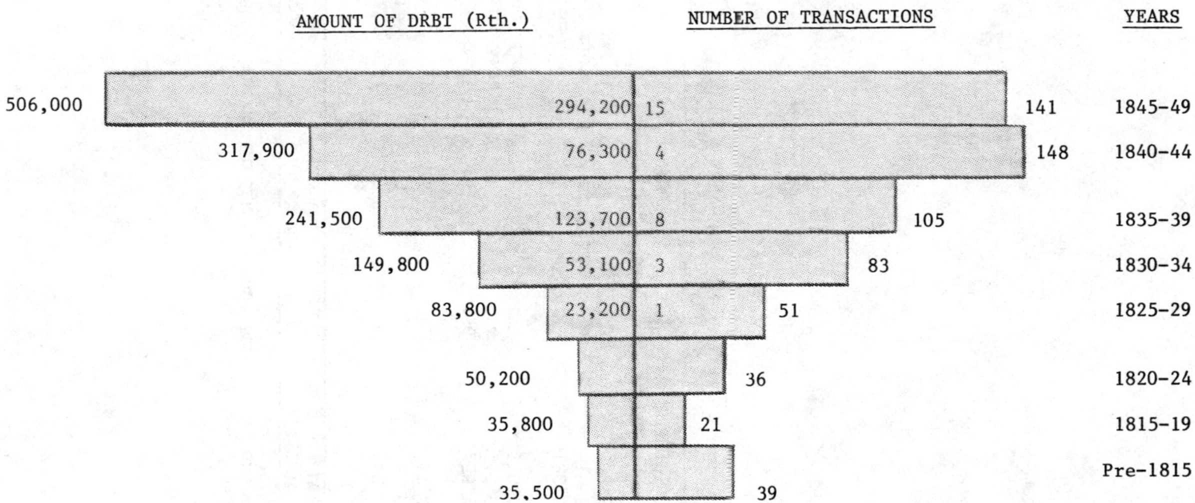

Figures within the shaded area represent the amount and number of transactions involving more than 10,000 Rth. per transaction.

Source: Debt Book

1790's to 1850, Owners, Creditors, and Amount of Debt by Residence

Area of Residence	Number of Owners	Creditors	Total Debt	Above 3,000
A. Ruhr valley	78	85	131,000	65,000
North of Valley	27	48	75,000	35,000
South of Valley	23	34	46,000	15,000
Unknown	16	10	17,000	10,000
B. North of Mining Region	14	56	255,000	232,000
South of Mining Region	35	104	210,000	155,000
East of Mining Region	15	33	150,000	135,000
C. Rhine Valley Region	2	22	188,000	176,000
Other German cities	1	8	15,000	10,000
Foreign	2	3	38,000	37,000
Total:	213	403	1,125,000	870,000

Source: Debt Book

1790's to 1850's Occupation of Mine Owners & Their Creditors

(A) Occupation of Mine Owners	Number (A)	Number (B)	(B) Occupation of Creditors (To Mine Owners)
I. Government Officials			
Mining Engineer	2	1	Mining Engineer
Mine Foreman	4	6	Mine Foreman
District Mine Manager	3	8	District Mine Manager
Mine Inspector	1	1	Mining Court Attorney
General Mine Official	2	11	Manager of Mine
Gvt. Accountant	1	1	Gvt.Acc.
Court Offical	2	1	Jr. Att. West.High Court
Jr. Gvt. Official	2	1	Senior Gvt. Lawyer
Court Clerk	1	6	Judge
President of Westphalia	1	1	Government Admin.
Tax Collector	1	1	President West.High Court
Official Head Forester	1	1	Ret.High Gvt. Official
Ret. Couty Assessor	1	1	Ret. Judge
Mayor	4	2	Mayor
Head of County	1	3	Head of County
Local Gvt. Att.	1	7	Local Att.
City Counselor	1		
Ruhr Tells Acc.	1		
City Collector	1		
II Professionals			
Colonel	2	1	Colonel
Lieutenant	4	5	Lieutenant
Medical Doctor	6	2	Medical Doctors
Surgeon	1	2	Surgeon
Lawyor	i	1	Lawyer
Druggist	i	1	Druggist
Estate Manager	i	3	Estate Manager
Teacher	4	9	Pensioner
		1	Law Student
		10	Preacher
III Landowners			
Estate Owner	6	3	Estate Owners
Nobels & Chamberlain	4	4	Nobles & Chamberlain
IV Craftsmen			
Smith	1	3	Blacksmith
Carpenter	2	2	Carpenter
Bookbinder	1	1	Bookbinder
Miner	7	5	Miner
DAy Worker	1	1	Dayworker
Distiller	1	1	Distiller
Farmer	6	11	Farmer
Clothprinter and Dyer	4	1	Cloth Dyer
San Stone Worker	1	1	Constructor of Wood Walks
River Transporter	1	1	Hammersmith
Steam Engine Mechanic	1	1	Mason
Shoemaker	4	1	Gold Craftsman
Hatmaker	1	1	Drilling Smith
Sissor Maker	1	1	Tanner

TABLE

IV. Craftsmen			
Coachman	1		
Brewer	1		
Kegmaker	1		
Butcher	2		
Clockmaker	1		
Stonecutter	1		
Wagon Driver	1		
Cartmaker	1		
Furniture Maker	1		
V. Commercial & Industrial Entrepreneurs			
Merchants	14	36	Merchant
Innkeeper	3	8	Innkeeper
Factory Owner	3	1	Manufacturer
Accountant	2	1	Director of Iron Works
Inventor	1	1	Machine Manufacturer
		16	Bankers & Banking Houses
		1	Steam Engine Manufacturer
		1	Store Owner
		1	Merchandising House
		4	
VI. Widows and Wives			
Widows	37	16	Widows
		1	Wife of Jr. Lawyer
		1	Wife of Head of Local Town
		1	Wife of Estate Owner
VII. Funds and Treasuries			
		1	Pension Fund
		1	Miners Fund
		2	West. Mine Aid Treasury
		4	Mining Office Treasury
		1	Ruhr Reparation Treasury
		2	Mark Mining Fund
		1	Court Ret. Fund
		3	B.A. Treasury
		1	Salary Pension Fund
		1	Poor Welfare Fund
VIII. Relatives		33	Relatives

Source: Debtors & Creditors

TABLE

1792, MINING OFFICE PERSONNEL UNDER VOM STEIN

Members of the Headquarters Office at Dertmund

1. Vom Stein	Director
2. Bordelius	Mining Judge
3. Mersbach	Senior Technical Official
4. Heintzmann	Senior Technical Official
5. Cappell	Accountant
6. Haardt	Accountant
7. Niemeyer	Surveyor
8. Kruener	Surveyor Help
9. Beck	Inspector
10. Thiel	Accounting Help
11. Schmeltzer	ccounting Help
12. Figge	Registration Clerk
13. Schmalenberg	Office Clerk
14. Deutecorn	Office Clerk
Also associated with the mining headquarters office:	
15. Liebrecht	Member of Provincial Government (War & Domain Authority)
16. Eversmann	Member of provincial government (War & Domain Authority)

Field Supervisory Personnel

17. Wuennenberg	Head of Mining District - Bachum
18. Crone	Head of Mining District - Hoerde
19. Coelln	Head of Mining District - Wetter
20. Engelhardt	Head of Mining District - Blankenstein
21-28 8 Officials	Technical Supervisors
29-36 8 Officials	Business Wupervisors

In addition to the above 36 officials directly employed by the mining office, there were by 1792 at least one person (if not 2) per mining operation who was employed by the mining office but paid for by the mining company. This individual (2 for larger operations) kept the journals. Nearly all the field officials had to inspect every operation at least ever 14 days.

Source: Reuss, Mitteilungen, pp. 327-330

TABLE

1838/39, MINING ADMINISTRATION PERSONNEL

A. Administrative Personnel*

Name (1)	Position (2)	Year-Present Position (3)	Age (4)	No. Children (5)	Place/Region of Birth (6)	Occupation of Father (7)
1. v. Derschau	Head of Region	1821	52	4	Sax.-Magd.	Colonel
2. Honigmann	Head-Essen M.O.	1814	63	8	Sax.-C.of Mans.	Innkeeper
3. Roehr	Administrator	1822	63	7	Westh.	M. District Head
4. Engelhardt	Administrator	1829	51	7	Westp.	Lawyer
5. Wiesner	Administrator	1831	35	2	Rhinel.	Royal Physician
6. Mummenhoff	Accountant	1834	43	5	Westp.-Boch.	Merchant
7. Anders	Secretary	1831	45	6	Silesia	M. Tech. Supv.
8. Juettner	Accountant	1822	49	5	Silesia-Bres.	Wood Merch.
9. Steinbrink	Clerk	1823	46	5	Rhinel.-Bres.	M. Accountant
10. Walter	Accountant	1830	47	6	Rhinel.-Stolb.	M. Bus. Mgr.
11. Messing	Accountant	1831	40	0	Westp.-Herd.	Preacher
12. Schalenberg	Accountant	1812	53	11	Westph.-Hern	M. Clerk
13. Bork	Head Clerk	1821	42	7	Westp.-Wet.	M. Accountant
14. Stephan	Clerk	1830	42	7	Westp.-Wet.	Revenue Supv.
15. Milde	Clerk	1833	40	0	Pol.-Raw.	Est. Mgr.
16. Grosche	Office Help	1827	48	1	Siles.-Neis.	Butcher
17. Brandt	MESSENGER	1828	56	4	Silesia	Soldier
18. Mahler	Messenger	1833	50	0	Pomm.-Stagrad	Farmer
19. Otto	Messenger	1836	45	6	Silesia	Carpenter
20. Hils	Supervisor	1837	35	0	Sax.-Nueh.	M. Dist. Head
21. V.D. Berken	Judge	1824	41	2	Westph.-Ket.	County,Cty.Jdg.
22. Bellmann	Lawyer	1833	33	0	Westph.	Merch.
23. Heine	Court Secretary	1822	47	5	Westph.	Merch.

* Most of the above personnel worked at the Mining Headquarters in Dortmund, the rest in the two regional offices in Essen and Bochum.

°M-Mining

Source: A-63

TABLE

1838/39, MINING ADMINISTRATION PERSONNEL IN MARK REGION:
B, Field Officials

(1) Name, (2) POSITION, (3) Year present Postion attained, (4) age,
(5) No. of Children, (6) Place of Birth, (7) Occupation of Father

(1)	(2)	(3)	(4)	(5)	(6)	(7)
1. Herold	D-H	1838	34	0	westp.	D.M. Engr.
2. Hardt	D-H	1817	47	6	SAX.	Ret. M. Off.
3. Crone	D-H	1822	46	10	WESTP.	Elder o.m. Un.
4. Reinhard	D-H	1830	39	0	L. Rh.	Preacher
5. V. Ende	D-H	1831	39	0	WESTP.	Merchant
6. Lind	D-H	1832	42	2	S-H	D-H
7. Jacobi	D-E	1833	37	2	Sil-Sch.	Merchatn
8. Bzettritz	D-E	1833	31	3	Bbndburg	Est. Owner
9. Kirchner	D-E	1838	26	0	WESTP.	M. Acctg.
10. Munscheid	D-E	1806	60	5	SAX.	M. Fore.
11. Vondran	D-E	1813	65	5	WESTP.-M	Miner
12. Schulze	D-E	1816	53	8	WESTP.	D.M.Engr.
13. Reiser	D-E	1833	43	4	RHN.DUS.	M. Tech. Supv.
14. Hilgenstck	D-E	1838	24	0	WESTPH.	D-H
15. Achtratt	D-E	1830	50	4	L. RH.	M. Foreman
17. Feldmann	D-E	1838	32	0	WESTP.	M. Accountant
18. Meyer	D-E	1839	24	0	WESTP.KETT	Merchant
19. Cremer	DOB	1832	37	2	WESTP	Baker Merchatn
20. Herberholz	D-B	1816	53	6	WESTP	M. Carpenter
21 Bergmann	D-B	1816	56	0	WESTP	Innkeeper,Frm.
22. Schmeltzer	D-B	1822	41	4	WESTP	Inkeeper
23. Landkuhl	D-B	1830	40	5	L. RH.	Clerk
24. Schroeder	D-B	1838	36	3	WESTP	M.B.Mgr.
25. Kaeseler	D-B	1837	31	0	WESTP	Miner
26. Wartesmann	D-B	1836	23	0	WESTP	D-H
27. Spahn	D-B	1835	37	2	WESTP	D-E
28. Frantzes	D-B	1837	31	0	WESTP	M. Foreman
29. Heckling	D-B	1815	42	4	WESTP	D-B
30. Schroeder	D-B	1834	38	4	WESTP	Estate Owner
31. Daniel	S	1829	38	4	WESTP	Miner
32. Fricke	S	1833	35	3	WESTP	Preacher
33. Meyer	M-E	1828	62	3	SAN-C. OF MSF.N/N	
34. Meinholt	M-H	1837	28	1	WESTP-KEON.	Slt. Wk. Direct.
35. Speer	M-H	1839	49	5	WESTP	Merchant.

* D-H Head of Mining District
D-E, District Engineer
D-B, District Bus. (Mgr.) Official
M-E, Machine Engineer
M-H, Help to Machine Engineer

S, Surveyor
M, Mining

Source: A-63

TABLE

1818-38, NUMBER OF MINING OFFICIALS

	1818	1826	1838
1. Personnal in the Mining Office	15	20	25
2. Medical Personnel	12	12	12
3. Business Official of Mining District	12	13	18
4. Technical Official of Mining District (Ingeniers	9	10	15
Total	48	55	70

Source: A-63

WEIGHTS & MEASURES

WEIGHTS

1 Zentner	= 100 Pfund (Metric System)
1 Metric Ton	= 20 Zentner = 2,000 Pfund = ca. 15 Scheffels = 12 Ringels = 4 old tons (ca. 21 Ztrs)
1 Ringel	= ca. 165 Pfund = 4400 cubic inches
1 Scheffel	= 1.35 Pfund (average) = 110 Pfund in 1850's = 1 English bushel = 3072 cubic inches
1 Malter	= ca. 4 Ringel = 4 Gang
1 Gang	= 1 Ringel
1 Old Ton	= 4 Scheffel = ca. 540 Pfund

MEASURES

1 Fuss	= 1 foot = 0.31 meter
1 Zoll	= 1 inch
1 Rute	= 3.7 meter
1 Lachter	= (1 Ltr.) = 2.09 meter
1 Maassen	= 28 Lachter
1 Ladung	= 33 Zolle = 2,000 Ztr. = 1 metric ton

UNITS OF MONEY

1 Old Reichsthaler (before 1815) = 60 Stüber = 2.50 Marks (1875)

1 Stüber = 12 Pfennige (before 1815)

1 Old Prussian Reichsthaler (before 1815) = 24 Gute Groschen
= 3.00 Mark (1875)

1 Guter Groschen = 12 Pfennige

1 Reichsthaler (after 1823) = 30 Silbergroschen (Sgr.) = 3.00 Mark (1875)

1 Sgr. = 12 Pfennige

DISSERTATIONS IN EUROPEAN ECONOMIC HISTORY

An Arno Press Collection

Atkin, John Michael. **British Overseas Investment, 1918-1931** (Doctoral Dissertation, University of London, 1968). 1977

Brosselin, Arlette. **Les Forêts De La Côte D'Or Au XIXème Siècle, et L'Utilisation De Leurs Produits** (Doctoral Thesis, Université de Dijon, 1973). 1977

Brumont, Francis. **La Bureba A L'Époque De Philippe II** (Doctoral Dissertation, Université de Toulouse, 1974). 1977

Cohen, Jon S. **Finance and Industrialization in Italy, 1894-1914** (Doctoral Dissertation, University of California, Berkeley, 1966). 1977

Dagneau, Jacques. **Les Agences Régionales Du Crédit Lyonnais, Années 1870-1914** (Doctoral Thesis, Université de Paris-VIII, 1975). 1977

Dennis, Kenneth G. **'Competition' in the History of Economic Thought** (Doctoral Dissertation, Oxford University, 1975). 1977

Desert, Gabriel. **Une Société Rurale Au XIXe Siècle:** Les Paysans Du Calvados, 1815-1895 (Doctoral Dissertation, Université de Paris, Sorbonne, 1971). 1977

Fierain, Jacques. **Les Raffineries De Sucre Des Ports En France:** XIXe -- début du XXe siècles (Doctoral Dissertation, Université de Nantes, 1974). 1977

Goreux, Louis-Marie. **Agricultural Productivity and Economic Development in France, 1852-1950** (Doctoral Dissertation, University of Chicago, 1955). With the Revised French Version. 1977

Guignet, Philippe. **Mines, Manufactures et Ouvriers Du Valenciennois Au XVIIIe Siècle** (Doctoral Dissertation, Université de Lille III, 1976). Two vols. in one. 1977

Haines, Michael R. **Economic-Demographic Interrelations in Developing Agricultural Regions:** A Case Study of Prussian Upper Silesia, 1840-1914 (Doctoral Dissertation, University of Pennsylvania, 1971). 1977

Hohorst, Gerd. **Wirtschaftswachstum Und Bevölkerungsentwicklung In Preussen 1816 Bis 1914** (Doctoral Dissertation, University of Münster, 1977). 1977

Huertas, Thomas Francis. **Economic Growth and Economic Policy in a Multinational Setting:** The Habsburg Monarchy, 1841-1865 (Doctoral Dissertation, University of Chicago, 1977). 1977

Jankowski, Manfred. **Public Policy in Industrial Growth:** The Case of the Early Ruhr Mining Region, 1766-1865 (Doctoral Dissertation, University of Wisconsin, 1969). 1977

Jefferys, James B. **Business Organisation in Great Britain, 1856-1914** (Doctoral Dissertation, University of London, 1938). 1977

Kirchhain, Günter. **Das Wachstum Der Deutschen Baumwollindustrie Im 19. Jahrhundert** (Doctoral Dissertation, University of Münster, 1973). 1977

Von Laer, Hermann. **Industrialisierung Und Qualität Der Arbeit Eine Bildungsökonomische Untersuchung Für Das 19. Jahrhundert** (Doctoral Dissertation, University of Münster, 1975). 1977

Lee, W. R. **Population Growth, Economic Development and Social Change in Bavaria, 1750-1850** (Revised Doctoral Dissertation, University of Oxford, 1972). 1977

LeVeen, E. Phillip. **British Slave Trade Suppression Policies, 1821-1865** (Doctoral Dissertation, University of Chicago, 1972). 1977

Metzer, Jacob. **Some Economic Aspects of Railroad Development in Tsarist Russia** (Doctoral Dissertation, University of Chicago, 1972). 1977

Moe, Thorvald. **Demographic Developments and Economic Growth in Norway, 1740-1940** (Doctoral Dissertation, Stanford University, 1970). 1977

Mueller, Reinhold C. **The Procuratori di San Marco and the Venetian Credit Market:** A Study of the Development of Credit and Banking in the Trecento (Doctoral Dissertation, Johns Hopkins University, 1969). 1977

Neuburger, Hugh. **German Banks and German Economic Growth from Unification to World War I** (Doctoral Dissertation, University of Chicago, 1974). 1977

Newell, William Henry. **Population Change and Agricultural Developments in Nineteenth Century France** (Doctoral Dissertation, University of Pennsylvania, 1971). 1977

Saly, Pierre. **La Politique Des Grands Travaux En France, 1929-1939** (Doctoral Dissertation, Université de Paris VIII, Vincennes, 1975). 1977

Shrimpton, Colin. **The Landed Society and the Farming Community of Essex in the Late Eighteenth and Early Nineteenth Centuries** (Doctoral Dissertation, Cambridge University, 1965). 1977

Tortella[-Casares], Gabriel. **Banking, Railroads, and Industry in Spain, 1829-1874** (Doctoral Dissertation, University of Wisconsin, 1972). 1977

Viallon, Jean-Baptiste. **La Croissance Agricole En France Et En Bourgogne De 1850 A Nos Jours** (Doctoral Dissertation, Université de Dijon, 1976). 1977

Villiers, Patrick. **Le Commerce Colonial Atlantique Et La Guerre D'Indépendance Des États Unis D'Amérique, 1778-1783** (Doctoral Dissertation, Université de Paris I, Pantheon-Sorbonne, 1975). 1977

Walters, R. H. **The Economic and Business History of the South Wales Steam Coal Industry, 1840-1914** (Doctoral Dissertation, Oxford University, 1975). 1977

DATE DUE

JUL 14 1981

GAYLORD